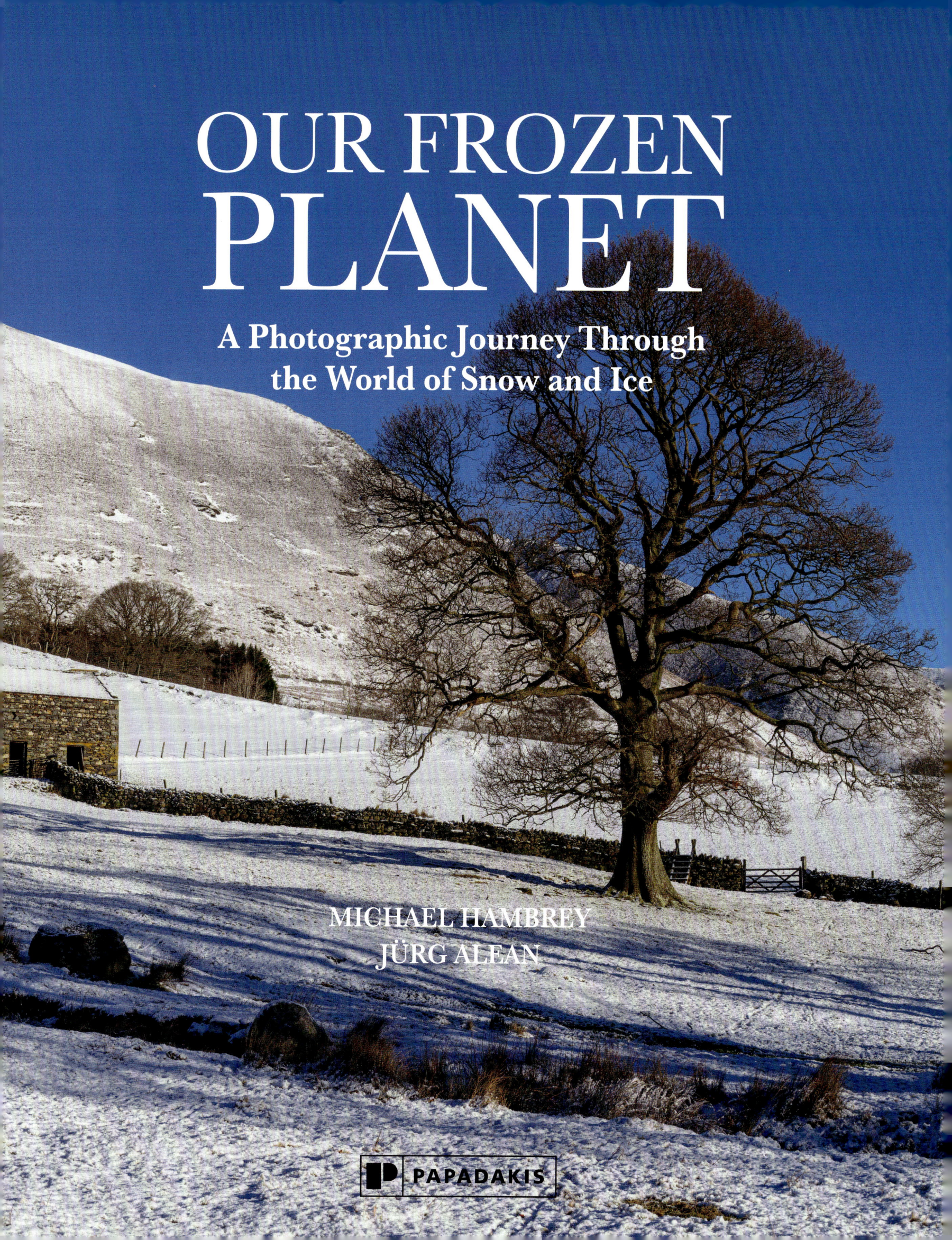

OUR FROZEN PLANET

A Photographic Journey Through the World of Snow and Ice

MICHAEL HAMBREY
JÜRG ALEAN

PAPADAKIS

*To glaciologists around the world who, through arduous field campaigns,
have alerted humanity to the causes and consequences of climate change.*

MICHAEL HAMBREY • JÜRG ALEAN

First published in Great Britain in 2024 by Papadakis Publisher

An imprint of Academy Editions Limited

Kimber Studio, Winterbourne, Berkshire, RG20 8AN, UK
info@papadakis.net | www.papadakis.net

@papadakisbooks

Publishing Director: Alexandra Papadakis
Design Director: Aldo Sampieri
Design & Editorial: Alexandra Papadakis
Proofreader & Indexing: Molly Dewar
Publishing Assistant: Tatiane Mauree

ISBN 978 1906506 735

Copyright © 2024 Michael Hambrey, Jürg Alean & Papadakis Publisher
All text and images © Michael Hambrey and Jürg Alean unless
otherwise stated.

No part of this publication may be reproduced or transmitted in any
form or by any means, electronic or mechanical, including photocopy,
recording or any other information storage and retrieval system, without
prior permission in writing from the Publisher.

All rights reserved.

Printed and bound in China.

Papadakis is committed to a sustainable future for our business, our
readers, and our planet. This book is made from Forest Stewardship
Council™ certified paper, responsibly sourced from forests that are
managed in an environmentally and socially responsible manner.

○ (Page 1) **Tokositna Glacier and Denali, Alaska, 2012**
Multiple debris stripes or medial moraines are typical of glaciers in the Alaska Range. Each moraine is formed at the confluence of two merging tributary glaciers. Denali (6144 m / 20,310 ft), North America's tallest mountain, provides the source for the ice in Tokositna Glacier.

○ (Page 2-3) **English Lake District, 2022**
At low levels in cool temperate regions, such as northwest Europe, snowfalls are increasingly sporadic and now rarely last for more than a few days. Nevertheless, mountain landscapes are transformed, as in this photograph of a stone barn, framed by two oak trees, near the village of Threlkeld.

CONTENTS

	Preface	6
1.	**The Cryosphere: An Introduction to the World of Snow & Ice**	8
2.	**Snow**	20
3.	**Ice From the Atmosphere**	36
4.	**Lake & River Ice**	44
5.	**Sea Ice**	58
6.	**Glacier Ice**	74
7.	**Disappearing Glaciers & Shrinking Ice Sheets**	100
8.	**Melting Glaciers**	122
9.	**Nature's Debris Conveyor**	144
10.	**Where Glaciers Meet the Sea**	158
11.	**The Legacy of Glaciers & Ice Sheets**	180
12.	**Life in a Frozen World**	202
13.	**Snow, Ice & Society**	228
14.	**A Farewell to Snow & Ice**	242
	Index	252
	Acknowledgements	257
	About the Authors	258

PREFACE

This book is a celebration of the world of snow and ice, or what scientists refer to as *The Cryosphere*. Although frozen water represents a significant component of Planet Earth, it is rapidly declining, both in area and volume, in response to rising global temperatures.

The investigation of snow and ice is a field of scientific endeavour that extends back several centuries in Iceland, Norway, and the European Alps. The science of glaciology evolved rapidly from the early nineteenth century onwards, when geologists and physicists took to mountaineering amongst the glaciers and snowfields of the Alps, and were drawn to study them. Deceptively simple in form, ice has proved a challenge to understand, so varied are its processes of formation and subsequent behaviour. As snow, ice crystals fall gently to the ground, light as a feather, with each snowflake being a unique and beautiful work of art. Snow flies through the air at the behest of the wind, or it flows downslope as a destructive avalanche. Ice crystals high in the atmosphere produce cirrus clouds that generate colourful optical effects such as haloes, and in cumulonimbus clouds produce hard-hitting hail stones.

On lakes and rivers ice forms intricate crystal growths, and may be as shiny as a mirror, or as translucent as milk. As river ice melts and breaks up, it forms a battering ram that can destroy woodland or damage human infrastructure.

Ice sheets preserve samples of the atmosphere in air bubbles for hundreds of thousands of years, giving us a remarkably detailed insight of past climatic history, and thereby laying the foundations for our future understanding of climatic change. In contrast, ice in glaciers can become incredibly contorted and is powerful enough to carve out majestic landscapes, eroding bedrock deep below sea-level. Snow and ice, so soft and delicate, yet also so powerful and destructive.

Snow and ice capture all our senses. Visually, ice changes colour and, while delicate blues predominate, reflected light produces all colours of the spectrum. Organic and mineral matter in the ice give rise to black, brown, orange, and grey colours. Orally, we can taste the pure freshness of ice or meltwater, and even enhance our drinks with air bubble-rich ice. The sound effects produced by ice are many and varied – from bangs, creaking, squeaking, and groaning as the ice fractures, to the tumultuous roar of a calving event where a glacier enters the sea or a lake. Even the smell of snow and ice is distinctive, as the associated low temperatures induce a fresh crispness in the air around us.

Global changes to snow and ice are now reinforcing our perceptions of climate change, and providing a stark warning of the impact of humans on our fragile planet. The language of climate change is fraught with political undertones, however, as elements of the fossil fuel industry and their political supporters have sought for decades to undermine the scientific consensus that humans are responsible for the warming of our planet. The authors' own perception that snow and ice are declining because of human-induced rising global temperatures is based on actual observations around the world. We believe that terms like 'climate change' and 'global warming' are too benign in relation to the situation in which we find ourselves. Thus, in this book, we make no apology for using the terms climate 'crisis' or 'emergency', and 'global heating', which are consistent with the beliefs of many environmentalists, aid agencies, conservationists, and an increasing number of nations, states, and local authorities.

The next few generations of humanity will see unprecedented change in snow and ice cover, so now is the time to place on record the magical beauty of such areas. The two authors have each studied glaciers for half a century, both as researchers and educators. We have had the opportunity to work in both polar regions, as well as several of the world's major mountain ranges. To this day, we are constantly struck by the beauty of these environments, and, in understanding their fragility, we have sought to document all forms of snow and ice, and their associated phenomena. We also recognise, with a degree of sadness and concern, that humanity has had, and continues to have, an adverse impact on the cryosphere. We therefore hope that this book will serve as a reminder of what we are losing, but also give us the courage to tackle the global climate emergency.

Michael Hambrey, Threlkeld, Cumbria, England
Jürg Alean, Eglisau, Kanton Zürich, Switzerland
October 2024

CHAPTER 1

THE CRYOSPHERE

AN INTRODUCTION TO THE WORLD OF SNOW & ICE

◀ Axel Heiberg Island, Nunavut, Canada, 2022
Glaciers, icebergs, lake ice and snow patches are all revealed in this image of Astro Lake, a glacier-dammed lake adjacent to Thompson Glacier (foreground).

We live on a planet where up to a third of the land area is covered by snow and ice during the winter, with up to a tenth of the land under 'permanent' ice. Taken together, snow, underground ice, sea ice, and glaciers all constitute the 'cryosphere'. This complements the other earthly 'spheres', such as the atmosphere, biosphere, hydrosphere, and geosphere. This book is intended as a celebration of the beauty of snow, river ice, lake ice, sea ice, and of glaciers and ice sheets. In it, we explore the processes involved in the formation of the different components of the cryosphere. We also consider how these components affect human civilisation, both positively and negatively. Throughout the book there is an underlying and unifying issue that humanity urgently needs to address – climate change, and how the cryosphere is responding to it.

Snowfalls induce mixed emotions in people. In the inhabited cooler parts of Earth, waking up after an overnight snowfall gives a feeling of awe at the beauty of snow-hidden landscapes, and many of us look with anticipation to venturing out in these conditions. Kids, especially, will want to take full advantage of these conditions, through sledging and snowball-fights. In Alpine and other snowy countries, people of all ages enjoy the skiing and snow-boarding opportunities offered by a well-developed tourist infrastructure. The opposing emotion evoked is perhaps one of dread, recalling memories of disruption to transport and the difficulties of getting to work, school, or college.

However, there is no doubt that there is something magical about a snowy landscape lit by a low winter sun. It is this totally fresh new appearance that gives us a sense of a minimalist reconfiguration of features that previously were familiar (Chapter 2).

Equally familiar to most people is the presence of ice in the atmosphere (Chapter 3), although it might not always be recognised as such. This ice comes in the form of clouds, ranging from the impressive towering cumulo-nimbus clouds that produce thunderstorms and hail, to the high-level thin veils across the sky that are a zone of ice particles. Humidity from the atmosphere also condenses on the ground or on plants, creating exquisitely delicate features known as rime and hoarfrost.

Ice on rivers, streams, lakes, and ponds may produce delicate or spectacular components of the landscape, especially when closely associated with snow (Chapter 4). In populated areas, we humans enjoy getting out onto lake ice if it is thick enough – for walking, skating or, if snow-covered, cross-country skiing and other activities. In the frozen North, such as the Yukon, Alaska and Siberia, temporary ice roads are the only surface means of getting around. However, more often than not ice is unstable and treacherous, and thus best avoided. On lakes and rivers in the more temperate regions the ice may simply be a thin film that lasts only a few days, but it can nevertheless be of considerable beauty.

Again, in populated parts of the world, usually in countries influenced by a stable cold continental climate in winter, it is common for marine coastal areas to freeze over. These areas then serve the same access opportunities as frozen rivers and lakes. It is, however, in the Arctic and Antarctic that we find the world's thickest and most extensive sea ice (Chapter 5). These beautiful areas are fragile, and, with their iconic animal species, are showing clear signs of sea ice decline, as an indirect consequence of human-induced global heating, which in turn is altering our weather patterns.

Lake and river ice, and snow, represent only a small percentage of the cryosphere on Earth's surface by volume. By contrast, glaciers and ice sheets represent the most important components of the cryosphere. However, by being distant from human settlement, they are relatively little-visited, except by mountain-lovers, skiers, and scientists. The formation and evolution of glaciers and ice sheets has long fascinated scientists, and their global impact is not to be under-estimated (Chapter 6). Glaciers are among the most stunning components of the landscape, offering a fascinating variety of features for the visitor to enjoy, or by which to be challenged. More importantly, glaciers and ice sheets, through changes in length and thickness, are among the best indicators of

▼ **Cadair Idris, Eryri (Snowdonia) National Park, Wales**
The glacier-carved cirque of Cwm Cau on the mountain Cadair Idris ("Idris' Chair"). Snowfalls now only occur sporadically during the winter, even at the higher elevations.

climate change. They provide one of the most vivid illustrations of how our planet is heating up at an unprecedented rate (Chapter 7). The great ice sheets over Antarctica and Greenland, and most of the world's glaciers, are shrinking, delivering meltwater to the oceans at an accelerating rate. Consequently, millions of people living close to sea level will be displaced, arable land will vanish, and cities will need to be relocated to higher ground. Furthermore, as mountain glaciers continue to disappear, as many more will do by the end of the century, the loss of water supplies and hydro-electric power generation capacity will have a severe impact on hundreds of millions of people.

Meltwater is among the most fascinating elements of a glacier. It creates a wide range of beautiful features, such as channels, canyons, holes, and tunnels (Chapter 8). Meltwater, when it reaches the glacier bed or the area beyond the ice front, is a powerful agent of erosion. Many glaciers also carry an enormous amount of debris, scouring the bedrock, and bulldozing sediment at their lower ends (Chapter 9). Many glaciers in polar and sub-polar regions, as well as the huge ice sheets of Antarctica and Greenland, terminate in the sea. The interaction between ice, ocean and atmosphere is complex, but there is no doubt about the magnificence of calving ice fronts and icebergs (Chapter 10).

Glaciers, through their capacity to erode bedrock and deposit sediment, are major sculptors of the landscape, thereby creating some of the world's finest scenery (Chapter 11). Since a third of Earth's land surface was covered by glaciers and ice sheets during previous ice ages, the last peaking only 20-30,000 years ago, their landscape legacy is profound.

Regions of snow and ice test the resilience of wildlife to the full, and reveal some outstanding examples of adaptation, whether it be penguins in Antarctica or polar bears and reindeer in the Arctic. Many species of birds and mammals rely on vegetation for food, and despite the ground

◥ **Luosto, Lapland, Arctic Finland**
Fresh snow and thick rime deposits on trees combine with a low sun to produce a magical winter wonderland that is well worth exploring on snowshoes.

◂ **Eglisau, Canton Zürich, Switzerland**
A circular 22 degree halo, together with an upper tangent arc, around the sun. This phenomena is caused by ice crystals within a uniform layer of cirrostratus clouds.

The Cryosphere

being frozen for much of the year, a variety of colourful plants blossom in summer, especially in Arctic, sub-Arctic, and alpine regions (Chapter 12). These plants sustain animals such as reindeer, musk oxen, arctic hares, and lemmings through the harsh winter months, even when snow covers the ground.

The resilience of humans is also tested in snowy and icy regions. This resilience, however, is heavily dependent on fossil fuels, a factor that is the prime contributor to global heating, and thus melting of the cryosphere. The presence of snow creates both hazards and enjoyment for humans, while glaciers are a major economic resource, providing water for hydro-electric power generation, irrigation, and drinking water, as well as generating opportunities for winter sports (Chapter 13).

Recent generations have benefited from the presence of snow, ice, and glaciers in so many ways, but our legacy has been, and continues to be, a slow destruction of these vital components of the natural world (Chapter 14). The cause is clear: rising temperatures brought about by adding greenhouse gases into the atmosphere through the burning of fossil fuel, destruction of ecosystems, forest clearance for livestock farming, and over-population. We have both the technology and the knowledge to slow, and ultimately reverse, this alarming state-of-affairs. Indeed, we must do so for the sake of future generations. However, the political will has been sadly lacking. So perhaps this book, in which we have tried to illustrate the beauty of the cryosphere, is for us a 'farewell to snow and ice'? We sincerely hope not, but only time will tell whether common sense will prevail over the deeply ingrained vested interests that care nothing for the natural world.

◀ **Axel Heiberg Island, Nunavut, Canada**
Caribou, like this male, are particularly well adapted to living in snowy conditions in the High-Arctic. Wide feet provide support on snow-covered terrain, and thick fur makes survival possible in extremely cold conditions.

▶ **Liverpool Land, East Greenland**
Polar bear on sea ice in Carlsberg Fjord.

○ (overleaf) **Canton Zürich, Switzerland**
Ice has formed around a waterfall near Gibswil. Icicles growing from the top, and ice formations growing from the bottom have coalesced into one continuous column.

▲ Wrangell Mountains, Southern Alaska, 2012
A meltwater stream meanders over the surface of Root Glacier.

◄ Morteratsch valley, Engadin, Switzerland, 2009
The well-monitored Swiss glacier, Vadret da Morteratsch, with Piz Bernina (4,048.6 m / 13,283 ft), viewed from the moraine formed in the Little Ice Age of the early 1800s. Since the photograph was taken, the glacier has receded above the valley floor.

The Cryosphere

CHAPTER 2

SNOW

▶ **Luosto, Lapland, Finland**
Recently fallen snow adorns boreal forest. Slender fir trees in the background can more easily handle the extra weight than the bushes in the foreground, which are on the verge of collapsing.

Snow crystals are among the most delicate objects in nature, but when they coalesce as snowflakes during precipitation, and then accumulate, consolidate, and turn into ice, they can create havoc. In mountain and polar areas, where snow turns to ice, one of the most powerful forces on Planet Earth is initiated. Snow is the solid form of water (H_2O) that crystallizes in the atmosphere. It can, temporarily, cover a quarter of Earth's land surface, and thus at times have a big impact on human civilisation. Snow can be a nuisance, causing much disruption. In contrast, snow is viewed by many as a beautiful, sparkling, almost magical, substance that can transform our landscape. Still, others enjoy snow for the sporting opportunities it offers, and consequently for many communities snow is a central component of their economy.

ORIGIN OF SNOW

From whence does snow originate? A snow crystal forms in the atmosphere by freezing of a droplet of 'supercooled' water, or water that is below the normal freezing point. The snow crystal grows around a nucleus, such as a particle of dust or a pollen grain. Once formed, the crystal grows further from additional water molecules as it falls through the air, clumping together to form a snowflake.

A single snowflake is formed of a crystal of ice that has a hexagonal (six-sided) symmetry and can grow to form an intricate network of outstanding beauty that nevertheless maintains that symmetry. As snowflakes fall, they often clump together. This is especially true when the temperature is only a little below freezing, since when snowflakes collide a little pressure melting and refreezing occurs, enabling them to coalesce. A practical example of this is that it is easier to make a snowball or a snowman from relatively warm snow than from dry, low-temperature snow.

It is often said that no two snowflakes are the same. Nevertheless, they can be grouped, depending on their overall appearance, into seven categories: plates, stellars, columns, needles, spatial dendrites, capped columns, and irregular crystals. The size and shape of the crystal depends on the air temperature and humidity, with snowflakes tending to grow larger at higher temperatures and humidity. Frozen precipitation also includes hail, which are hard spheres of ice, and is described in Chapter 3. An in-between state consists of granular snow pellets called 'graupel' or soft hail. When snowflakes are melting and mixed with raindrops the precipitation is referred to as sleet. At very low temperatures of −50 °C (−58 °F), a common feature of the Arctic and Antarctic atmosphere is 'diamond dust', which consists of tiny ice columns and needles which float to the ground gently, even when the sky is clear and the sun shining.

The structure of cold fresh snow, where crystals have many branches, is fragile. When people walk across slightly consolidated snow these branches break and rub against each other, which makes the familiar crunching sound. In temperatures well below freezing, walking across wind-hardened snow produces a squeaking sound. When the snow is close to the melting point, the crystal branches bend and cease to make these sounds.

Snow accumulates intermittently during the winter months, and in colder regions may build up into a snowpack several metres thick. In regions of high elevation and in the polar regions, some of the snow survives from one summer to the next and multiple layers build up. The accumulation of multiple years of snow creates a glacier through recrystallisation into ice (Chapter 6).

SNOW-GENERATING WEATHER

We all love to talk about weather. This is especially true when depressions (low-pressure systems) sweep across the land. It is the winter depressions that are responsible for most snowfalls, but snow can also fall in spring and autumn. In areas under the influence of maritime conditions, such as northwest Europe, snowfalls may be associated with the passage of a cold front, the leading edge of a colder mass of air than the air in front. Cold and warm fronts may follow in quick succession, resulting in freezing and thawing. Snow may thus be ephemeral and fail to build up to any degree, much to the frustration of those who use snow for recreation. A warm front can also generate snowfall, when warm moist air flows over the top of sub-zero air at ground level, conditions which are more common in areas of continental climate such as the interior of North America. Snowfall is also influenced by topography such as lakes and mountains. Lake-induced snow is the

▸ *Although, on a microscopic scale, each snow crystal is unique, it shares a hexagonal (six-sided) symmetry with all others. Crystals are mainly 0.5–1cm in diameter. Photograph courtesy of Ulrike Sobick.*

result of cold air moving over warmer, moister air over the lake. Belts prone to snowfall occur, for example around the Great Lakes in North America. Mountain ranges also facilitate the accumulation of snow, since moist air hitting an elevated area rises and cools, releasing its moisture as snow. The snowline may fluctuate with altitude dramatically as weather fronts pass over the mountains.

Meteorologists talk about snow deposition in terms of intensity. From least to greatest, these events are referred to as snow flurries, snow showers, snowstorms, and blizzards. The UK Met Office definition of a blizzard is of moderate or heavy falling snow with winds speeds of 50 km/hr (30 mph) or more, and a reasonably extensive snow cover reducing visibility to 200 metres (656 ft) or less. Formal definitions are similar in North America and elsewhere, but many people use the term 'blizzard' more loosely. For example, a person may experience a ground blizzard where their head is mainly above a blowing mass of snow. An extreme condition of a blizzard is known as a white-out, but this term is also applied to snow and fog where ground and sky are indistinguishable. With blizzards and snowfalls, the media in countries that rarely get heavy snowfalls can get quite hysterical about the impact on humans. For example, terms such as *Snowmageddon* and *Beast from the East* have appeared in bold black letters in British tabloid newspapers, when perhaps only a few centimetres of snow have fallen!

▲ **Saariselka, Lapland, Finland**
The region is commonly regarded as home to Santa Claus or Father Christmas, from whence he distributes presents to children around the world.

▶ **Saariselka, Lapland, Finland**
A wooden hut on a cross-country ski trail carries a thick snow cover. The layers are a result of successive snowfall events.

THREATS FROM SNOWFALL

Large accumulations of snow on steep mountain slopes result in a significant threat to human life – avalanches. These rapid flows of snow are usually triggered on slopes with angles ranging from 30 to 45 degrees, given the right conditions. They can also run out for considerable distances over much more gentle ground. Where snow exists, avalanches may occur at any time of the year but, in the Northern Hemisphere, are most common from December to April. The likelihood of an avalanche taking place is dependent on several factors. These include the weather, the depth and temperature profile of the snow, snow structure, slope steepness and orientation, effect of wind, and the presence or absence of vegetation. Some of these conditions, such as temperature of the snowpack, can change on a daily or hourly basis, making avalanche prediction a serious challenge. Assessing avalanche risk is an important discipline in populated areas.

Snow scientists have developed a three-fold classification of avalanches. A powder snow avalanche comprises dry fresh powder-snow, commonly overlying a dense avalanche. Avalanches of this type travel very fast (up to 300 km/hr; 190 mph), and flow over long distances, including uphill at their leading edge. A slab avalanche occurs in wind-deposited snow and is characterised by fracturing of the snowpack and displacement as a block, which then disintegrates. A wet snow avalanche is a mix of snow and water moving downslope in suspension. Such avalanches are slow-moving (typically 10-40 km/h; 6-25 mph), since there is a lot of friction at the base of the flow. However, their high density means that they can also be very destructive.

↖ **McMurdo Ice Shelf, Ross Sea, Antarctica, 2010**
During blizzards visibility is reduced to a few metres, leading to 'white-out' conditions where the ground cannot be distinguished from air. An oversnow vehicle, the Swedish-built Hägglunds, is being checked during a blizzard.

▸ **Troms region, northern Norway**
Drifting snow, a so-called ground-blizzard, challenges drivers on a wintery road.

◂ **Weissfluhjoch, Davos, Switzerland**
A block of snow, about 1 m tall, extracted from the snowpack above the famous resort of Davos. The stratification or layers visible signify multiple snowfall events, and their documentation allows us to predict the avalanche risk in the area.

EFFECT OF WIND ON SNOW

Let us now turn to the many effects of wind on snow. We can view wind as a snow-sculptor, since it can create a wide variety of interesting and beautiful phenomena. At the lower end of the wind scale, the steady flow of wind over a loose snow surface can produce ripples that are reminiscent of those on sandy river-beds and beaches. Larger still are snow dunes which, just like their terrestrial counterparts, show cross-bedding, formed when snow is deposited on a lee slope or behind an obstacle. Wind may also erode the snowpack, producing ridges and furrows known as 'sastrugi', a word of Russian origin. Wind-compacted snow is quite hard and firm to walk on. However, traversing heavily wind-carved terrain full of large sastrugi is arduous and tough on transport equipment. In snowbanks influenced by gentle winds and sunshine, snow scallops may form. These are large groups of dish-shaped depressions, typically 10-20 cm in diameter, caused by small eddies in air currents and variations of sun angle and strength.

In areas where snow lies for most of the winter, spring melt is a major event. Large volumes of meltwater rush into rivers, which are often over-topped, resulting in widespread flooding. During such flood events erosion of the landscape is intense. A typical feature of a mountain environment with extensive winter snow accumulation is a 'braided river', reflecting huge variations of discharge during the year.

GLOBAL EXTENT OF SNOW

Using satellites, the extent of snow has been monitored over the Northern Hemisphere by the United States' National Oceanic and Atmospheric Administration (NOAA) from 1966 to the present-day. The snowiest month by area covered is January, when on average there has been 47 million square kilometres (18 million square miles), roughly twice the surface area of the North American continent. Snow cover is most extensive in the Arctic, where it covers the ground for up to nine months of the year (and all-year round on the upper reaches of glaciers, ice caps and the Greenland Ice Sheet).

In terms of areal extent, but not volume, snow is the most important part of the cryosphere. It has a huge influence on weather and climate. It keeps Earth's surface cool, because it reflects most of the incoming solar radiation back into space. Over the last half-century, the period monitored by NOAA, there has been an overall decline in snow over the Northern Hemisphere. On the one hand, it remains on the ground for shorter periods of time, and, on the other, the areas covered by snow have become smaller. The rate of decline has even accelerated since 2003. This trend is a consequence of global heating. Also playing a part in snow reduction is the impact of pollutants settling out from the atmosphere, such as increasing soot and dust from human activities, which darken the surface and aid melting. In many parts of the world this reduction in snow cover and melting is increasingly responsible for water shortages.

▶ **Walenstadt, Canton St Gallen, Switzerland**
Explosives were used to trigger an avalanche near Walenstadt. Artificial release of snow avalanches is used to avoid further accumulation of snow and, potentially, much bigger and, therefore, destructive avalanches.

○ **Castlerigg, Keswick, English Lake District**
The Helvellyn range (reaching 950 m / 3,118 ft in altitude) with a sharply defined snowline at around 600 m, viewed from the Neolithic (3,300 to 900 BC) stone circle of Castlerigg.

Snow

◐ **Bryce Canyon, Utah, USA**
Slope aspect strongly affects the snow distribution in mountain terrain. Almost all the snow has melted on the south-facing slope of this ridge, whereas a substantial cover remains on the shady, north-facing slope on the left.

Textures on the surface of snowpacks, formed by wind and sun.

◂◂ **Minna Bluff, Ross Sea, Antarctica**
Snow scallops or suncups formed by strong solar radiation and air eddies.

◂ **Near Scott Base, Antarctica**
Snow dunes with cross-bedding.

▴ **Blencathra, English Lake District**
Cornices overhanging the cliff-edge of the 800m-high mountain. They were formed during an unusually heavy snowfall event, combined with a strong northerly wind.

◥ **Near Saariselkä, northern Finland**
Snow ripples above the tree-line.

▸ **McMurdo Ice Shelf, Antarctica**
Wind-etched snow or sastrugi.

◢ **Blease Fell, Blencathra, English Lake District**
Sastrugi and cornices, with Ruby the border collie for scale.

Snow   35

CHAPTER 3

ICE FROM THE ATMOSPHERE

◀ **Near Sodankylä, Finland**
Plate-shaped ice crystals floating horizontally in the air act as countless tiny mirrors, reflecting light from a source on the ground, producing an 'ice pillar'.

Most snow and ice on our planet originate from moisture in the atmosphere, which in turn is derived from evaporation of the oceans, rivers, and lakes. Together they constitute the 'water cycle', the lifeblood of our planet. Because of condensation of moisture, clouds form which, with their multitude of shapes, adorn our skies. If temperatures are well below freezing and atmospheric condensation is sufficiently intense, we will experience snow, sleet, hail, frost, or rime, all of which are considered solid forms of precipitation. These forms are responsible for remarkably colourful phenomena in the atmosphere and provide features of outstanding beauty on plants and on the ground. Here we take a closer look at several types of ice of atmospheric origin, with the exception of snow.

FORMATION AND APPEARANCE OF ATMOSPHERIC ICE

Normally, on Earth's surface, water freezes at zero degrees Celsius. In fact, the Celsius temperature scale was originally defined using the freezing and boiling points of water. However, in the atmosphere conditions are somewhat different. The starting point for ice crystal formation is an impurity in the atmosphere, such as an aerosol or dust particle. The particle provides a nucleus around which water vapour can condense. Water droplets in clouds may become much colder than 0° C without freezing, hence the term 'supercooled water'. Ultimately, especially if disturbed by wind, the water droplets will freeze. In low clouds, where temperatures are only a few degrees below freezing, most water is present in the supercooled form, whereas in higher and colder clouds, a greater proportion of water has frozen and is suspended in the air in the form of tiny icy crystals. Like cloud droplets, these tiny ice crystals are so light in weight that they may remain airborne for an almost indefinite time, in a similar manner to dust particles.

Clouds consisting primarily of ice crystals are often thin, transparent, and have a feathery appearance. They are known as cirrus clouds (from the Latin *cirrus*, feather). Somewhat thicker and more continuous ice-clouds are called cirrostratus, which cover most of the sky with a semi-transparent, unstructured layer.

Often, when the sun shines through cirrus or cirrostratus clouds, beautiful optical effects may be seen. Most common is a large circular halo that surrounds the sun. It is reminiscent of a rainbow, but the colours typically are less prominent. Rainbows, in contrast, are observed with one's back to the sun, and are produced by both refraction and reflection of light inside spherical water droplets. Haloes are produced in similar clouds, but only by refraction inside tiny ice crystals; they are best seen when a tree or pole blocks the glare from the sun. Sometimes, particularly bright spots, referred to as 'sun dogs', are embedded within the halo to the left and right of the sun, as well as above it. This effect is formally known as a 'parhelion'. Occasionally, these sun dogs are astonishingly bright and consequently are also sometimes called mock suns. Haloes and sun dogs are a clear sign of ice crystals in the air in temperate or tropical latitudes, because even there the temperature in the high atmosphere is many degrees below zero. In high latitudes they also form within an icy haze much closer to the ground.

Other clouds containing ice are the towering cumulonimbus clouds, sometimes referred to as anvil-shaped clouds, which are associated with thunderstorms. They commonly build out of billowing cumulus clouds and contain not only ice but also many supercooled water droplets. An aircraft flying through such clouds provides surfaces on which the water vapour immediately freezes, forming a thickening crust of ice. This in turn changes the aerodynamic shape of the fuselage and propellers, blocks air intakes, and generally increases the weight of the aircraft, unless it is equipped with appropriate de-icing equipment. Cumulus and cumulonimbus clouds are always associated with strong thermal updrafts. Air rises within these clouds, cools with increasing altitude, causing more and more water vapour to condense resulting in cloud droplets and, ultimately, ice crystals. Once an ice crystal has become large and heavy enough it starts to fall despite the updraft. During its fall, it grows in size due to collisions with supercooled water droplets. The falling ice grain may subsequently end up in an area of even stronger updraft, which will lift it into higher regions yet again. Sometimes ice grains go through dozens of cycles of falling and rising, eventually leading to the formation of hailstones the size of tennis balls or even larger.

▸ **Bäretswil, Canton Zürich, Switzerland**
Cirrus clouds are made from ice crystals. These refract sunlight and sometimes lead to complex and colourful optical displays in the atmosphere, such as this halo around the sun, together with a second, coloured arch nearer the horizon.

Freezing rain is another phenomenon that involves supercooled water. If snow falls from freezing clouds into relatively warm air it turns to rain, but if a further mass of freezing air exists near ground level, the rain becomes supercooled. When the raindrops hit an object or the ground they immediately freeze, producing conditions that are treacherous to humans. Power lines may be brought down due to the weight of the ice; vegetation, buildings, and vehicles may be coated with ice; and roads become impassable. Conditions like this are not unusual in North America, but in Europe they are quite rare.

RIME AND HOARFROST

Another spectacular effect of supercooled water is 'rime'. Not to be confused with hoarfrost (see below), it comes in either of two forms: hard and soft. Hard rime ice is deposited in freezing fog during windy conditions when millions of supercooled water droplets freeze upon impact with an obstacle and build up an icy crust, usually on the windward side of objects. Rime-formation may continue over several days until a thick cover of ice has built up. Trees encased in such 'rime-armour' eventually will no longer resemble trees, but rather a myriad of shapes that stretches our imagination. Northern Scandinavia is particularly well known for landscapes adorned with rime-clad trees, usually found at moderately high altitudes, just below the treeline, where a layer of fog is common and facilitates the deposition of rime. Rime is also a feature of temperate mountain regions, such as in the UK, where it coats everything from rocks to vegetation and fences, forming delicate feathery patterns.

In contrast to rime, the formation of hoarfrost occurs during calm and clear weather. At night, when clouds are absent and heat-loss into space is strong, air temperatures near the ground drop the most. Sometimes, the air temperature near the ground may be much lower than a few metres, or even a few hundred metres, higher up. This effect is referred to as a temperature inversion, meaning that

▲▲ **Rovaniemi, Finland**
Halos within cirrus clouds sometimes are associated with very bright spots, so-called sun-dogs, at the same altitude as the sun.

▲ **Eglisau, Canton Zürich, Switzerland**
During a thunderstorm, hail has fallen on the tiled roof of a house. These hailstones are less than one centimetre in diameter. However, some hailstones can be bigger than tennis balls and cause considerable damage to crops, vehicles or even roofing.

▲ Scafell Pike, English Lake District
Rime formed from mountain mist during windy conditions at about 970 m / 3180 ft on England's highest mountain, Scafell Pike. The radial pattern of ice crystals is a consequence of air turbulence around the boulders. The largest boulder in the image is about half-a-metre (1.5 ft) across.

Ice From The Atmosphere

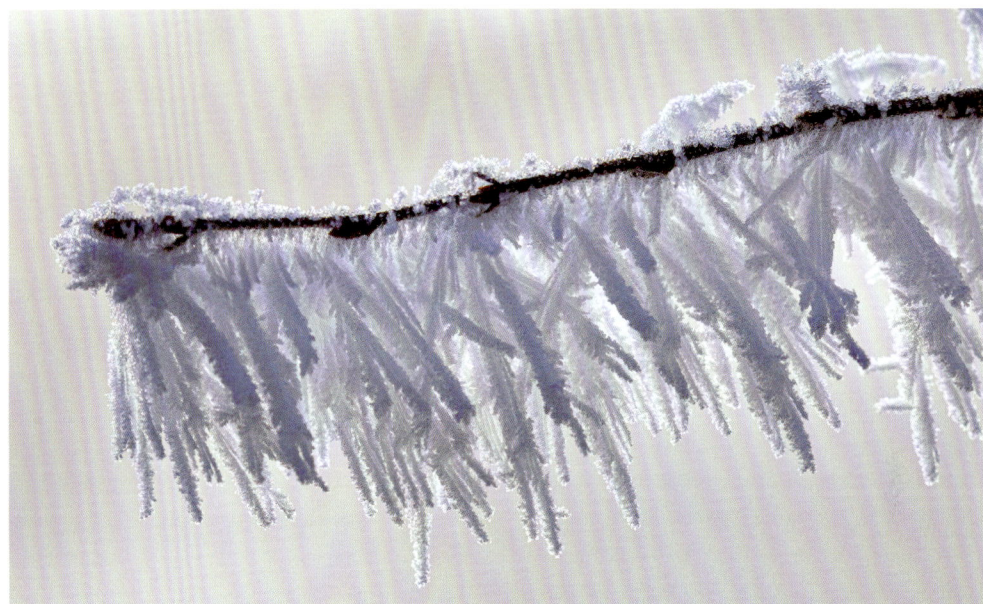

within the inversion temperatures increase rather than decrease with altitude. Strong inversions and high humidity are the ideal conditions for creating hoarfrost. As air temperature drops, the air can no longer hold all its water vapour. The excess is then deposited on rocks, grasses, bushes or trees. In the absence of wind, crystallisation of ice takes place slowly. Water molecules arrange themselves in a highly ordered fashion, forming delicate and highly complex crystal shapes, but all with the standard hexagonal structure. While often only a few millimetres in size, larger crystals several centimetres long may also form. On a sunny morning, frost-covered grasses and trees may briefly resemble a winter wonderland, until temperatures rise above the melting point or a breeze puts an end to the delicate icy spectacle. The term "hoar" derives from Old English, where hoar was defined as "showing signs of old age" in reference to old men with white beards.

▲▲ **Griessen, Baden-Württemberg, Germany**
Small crystals of hoarfrost have grown overnight on some wild raspberries.

▲ **Hemmental, Switzerland**
Crystals of hoarfrost, 1-2 cm (1") long, have grown on this twig.

◀ **Luosto, Lapland, Finland**
Thick crusts of rime have been deposited during moist and windy conditions on a group of fir trees.
The presence of supercooled atmospheric water droplets, the source of the rime, is indicated by the rainbow.

Ice From The Atmosphere   43

CHAPTER 4

LAKE
&
RIVER ICE

▶ **Fischenthal, Switzerland**
Spray from a waterfall has produced a curtain of icicles during an increasingly rare period of severe frost.

Fresh water ice is part of the human experience, forming annually on about half of Earth's surface waters in the Northern Hemisphere, primarily on lakes and rivers. Ice-formation may last only a night or two in warmer climates, whereas in frigid areas, such as Antarctica, some lakes are frozen all-year-round. As with other components of the cryosphere, ice in lakes and rivers has been showing a general decline in recent decades, in response to global heating.

LAKE ICE

Most lakes consist of layers of water at different temperatures, which we refer to as 'thermal stratification'. In summer, relatively dense cold water occurs at depth, and is overlain by an increasingly less dense layer as summer progresses. Heading into winter, the surface water cools to the same temperature as the deeper water, and then begins to mix with the water below. When the water temperature drops to 4°C (39°F), the maximum density of water is reached. Cooling beyond this point lowers density, and a surface layer is formed that begins to freeze when the temperature reaches the freezing point (0°C / 32°F). The fact that ice floats on its own liquid makes it an almost unique substance, and if this were not the case life as we know it could not exist.

The formation of ice is by a process known as 'nucleation'. Near the freezing point, ice formation requires tiny particles acting as nuclei to trigger the freezing. The first signs of freezing are often found at the lake shoreline, where the water is shallow and cools faster. The freezing zone then spreads out across the lake. A thin film of 'black ice' forms across wide areas of a lake in calm sub-freezing conditions. After reaching a certain thickness, black ice on lakes often produces the most unusual and surprisingly loud, eerie, metallic noises that travel back and forth over the lake. This strange effect results from the release of tension within the ice cover. This tension is caused by lateral expansion and contraction of the ice as the temperature changes between day and night.

When conditions are windy, formation of a continuous ice cover may be prevented. Instead, ice crystals gather into floes that constantly break-up and reform, until the wind lessens in strength. Floes first jostle each other, forming pressure ridges where they collide, until freezing causes them to lock-down into a continuous sheet.

In regions where the lake ice cover lasts the entire winter season, the decay of ice in spring takes place by thinning, rotting, and melting. The warmth of the sun causes surface melting and the ice becomes grey and blotchy, with surface pools forming that eventually connect with lake water beneath. The pools are dark and enhance the melting by absorbing more solar radiation than the lighter coloured ice. Melting is first evident near the shoreline, forming a 'moat', and eventually the lake ice breaks down into individual floes, following which the lake rapidly become ice-free.

The continental parts of North America have many large lakes that are prone to freezing. Along the border between Canada and the USA, the Great Lakes largely froze over until the early 1970s, but since then ice cover has only reached 55% on average, but with large variability between years. A consequence of a reduced ice cover on the lakes has been more moisture in the atmosphere, so the surrounding lands are receiving heavier snowfalls in winter. However, large lakes further north, such as Great Bear Lake and Great Slave Lake, still freeze over completely. In northern Europe, especially in Finland and Siberia, the lakes are much smaller, but they, too, experience wide variability of freeze-up and melting dates.

There are some highly unusual, but very interesting lakes in the polar regions. Saline lakes occur in the dry polar deserts of Antarctica, such as the Victoria Land Dry Valleys. There, several lakes are fed by meltwater from glaciers, but without outlets, salts can build up. These saline lakes have a permanent ice cover, with salts precipitating out at the surface. The water at depth remains liquid, since below 4°C (39°F) water expands thermally and so becomes less dense, while protected from further chilling by the surface ice layer. These Dry Valleys lakes are a unique example of strong thermal stratification, with minimal mixing between near-surface and deep water, since wind and wave action have little influence in causing turbulence in the water. Another unusual feature in Antarctica is the presence of permanently frozen 'epi-shelf' lakes, where water bodies are held back by a floating ice shelf, and have a connection to the open sea underneath. Lakes that freeze over every winter are also associated with glaciers elsewhere, but these are treated in Chapter 8.

▶ **Lej da Silvaplanau, Engadin, Switzerland**
High-altitude basins, such as in the Engadin, sometimes experience extremely low temperatures during calm weather conditions. After a few days, a layer of 'black ice' has formed, covering the whole of the lake. The reflection is not on water, but on the lake's black-ice surface.

RIVER ICE

Over a third of Earth's land surface is drained by rivers that freeze over at some stage during the winter. Seasonal ice influences how a river flows, which in turn affects the surrounding ecosystems, as well as having consequences for people and infrastructure, such as navigation obstruction for boats, or the use of frozen rivers as ice roads, as in northern USA, Canada and the Russian Arctic.

Ice-formation in rivers is affected by the velocity and turbulence of the water, and so is a more complex process than in lakes. Because of mixing, cooling is uniform through the water column, and freezing occurs when the water is supercooled to a temperature of just under 0°C. The nucleation of ice crystals is aided by grains of snow falling into the river, by sediment or organic particles carried by flow, and by motion of the water. The ice often forms as tiny crystals (less than a millimetre in size), often having the form of small discs – referred to as 'frazil ice'. These ice crystals coalesce to form slushy opaque ice or irregular masses of ice called 'pans'. Frazil ice can also form on the bed of a stream or river, where it is referred to as 'anchor ice'.

The formation and expansion of an ice cover begins along the banks of a stream or river. The loose ice pans merge forming larger floes, eventually freezing over completely. An ice cover has the effect of slowing the flow of the river and reducing discharge. Once the ice cover has become stable, it slowly thickens if temperatures remain below the freezing point. Growth of ice crystals sees them aligned vertically in interlocking columns.

○ **Coire Ardair, Grampian Highlands, Scotland**
In some years, the Scottish Highlands hold snow well into the spring, as seen here draping the glacially carved rock basin on Creag Meagaidh (1,130 m / 3,710 ft). Coire Ardair is one of the finest corries or cirques in the United Kingdom.

▶ **Lej da Silvaplanau, Engadin, Switzerland**
Methane released by biological activity at the bottom of the lake collects under the 'black ice'. As the ice hickens, the gas bubbles become trapped progressively further down, resulting in vertical 'bubble chains'.

▼ **Pfäffikersee, Canton Zürich, Switzerland**
Growth of complex ice crystals forms an intricate pattern near the shore of this lake. Rarely, pedestrians and skaters may cross the lake, once the authorities have declared the ice-cover thick enough and, therefore, safe.

○ **Eglisau, Canton Zürich, Switzerland**
Ice floes on the River Rhine (Rhein) near one of the author's home, during a spell of unusually low winter temperatures. A group of adolescent swans swimming up-river does well to avoid them.

By the time of the spring melt, the warming water causes the fixed ice cover to melt and disintegrate into separate pieces. These are then carried downstream, and if the channel has a constriction, or becomes shallower and less steep, 'ice-jams' may occur. The damming effect on water-flow may be sufficient to induce a flood event, resulting in damage to property and infrastructure. Those rivers most prone to ice jams are those that flow from south to north, or from warmer to colder climes, such as those in western Canada and Siberia. In those rivers, the ice in the headwaters melts first and is then free to travel downstream where it meets fixed ice.

In rivers with multiple channels, referred to as 'braided rivers', sheets of ice form that last well into the summer. This ice forms when water-flow continues during the initial freeze-up of early winter. It is known as *Aufeis* (from the German), or as 'icings'. The water flows from lakes, springs or glaciers, forming a sheet-like mass of layered ice, with each layer made up of vertically orientated crystals, sometimes referred to as 'candle ice'. When weathered, the crystals may reach about 25 cm or a foot in length and have an exquisite beauty when sheets of them collapse. *Aufeis* is common in High-Arctic regions, slowly disintegrating in summer as active river channels melt the ice from top-down and from underneath.

The above processes refer primarily to larger rivers in lowland areas or in wide, open valleys. In contrast, freezing of mountain streams and fast-flowing rivers creates many beautiful forms, especially when water flows over cliffs producing large icicles. Associated with the turbulence of fast-flowing streams, we can also expect to see frozen spray, which coats stones and vegetation alike.

Remarkable ice formations also form when water enters a cave in which sub-freezing temperatures are maintained for most of the year. Such cold caves can be found even outside permafrost areas. If the cave system has a tall entrance, cold winter air flows inside and, because of its relatively high density, will remain trapped during the summer months. Caves with massive icicles and ice stalagmites have been developed as major tourist attractions in Austria and Slovakia.

▲ Ablation Lake, Alexander Island, Antarctic Peninsula
An epishelf lake in Antarctica connected beneath the George VI Ice Shelf to the open sea. The surface of the lake is permanently frozen, but forward movement of the ice shelf into the lake basin produces pressure ridges along the shoreline.

▼ Dobšinská ľadová janskyňa, Slovakia
Part of a large limestone cave system in northern Slovakia, this 'ice cave' is partially filled by a giant body of water ice with a surface area of nearly 10,000 square metres (11,960 square yards), and a volume of more than 110,000 cubic metres (3,884,613 cubic feet). The cave entrance is at an elevation of 969 m (3179 ft).

▲ Fountain Glacier, Bylot Island, Canada, 2014
Close-up view of disintegrating Aufeis at Fountain Glacier, showing the exquisite form of the component candle-like crystals, each approximately 25 cm (1 ft) long.

◯ Derwentwater, English Lake District
Intricate patterns in frozen lake water at the edge of the lake, showing successive freezing fronts growing out from the shoreline and around stones.

CLIMATE CHANGE

The extent of winter lake and river ice has declined globally, especially in the last three decades – the period for which repeat satellite data are available. This is the inevitable consequence of the warming climate, which has the effect of delaying the winter freeze-up and hastening its thaw. The existence of ice cover and the timing of break-up of ice on lakes and rivers is of vital ecological importance. However, the duration and thickness of ice cover have been declining in the northern hemisphere, especially in Eastern Europe, the Tibetan Plateau, and Alaska. Future projections indicate further decline in the duration and extent of ice cover in lakes and rivers.

CHAPTER 5

SEA
ICE

◀ **Approaching the North Pole, 2019**
The Russian nuclear-powered icebreaker icebreaker "50 Let Pobedy" ("50 Years of Victory", referring to the anniversary of victory of the Soviet Union in World War II), carving a path through multi-year ice close to the North Pole. Warming of the region is indicated by pools of meltwater on the ice (pale blue), and areas of open water (black) between some of the floes. The vessel was charted by a polar cruise ship company, allowing guests to reach the Pole in comfort, as well as gaining an overview of the sea ice from a helicopter.

Sea ice forms by the freezing of ocean water, and in some areas complements iceberg ice which has calved from glaciers flowing off the land. Although it is found mainly in the Arctic and Antarctic, sea ice also forms in enclosed bays and seas to the south of the Arctic Circle, as well as around the coasts of temperate regions in very severe winters. Polar sea ice retreats in summer but does not entirely disappear. However, it is shrinking dramatically in terms of thickness and extent, and these changes are having a profound effect on polar ecosystems, ocean circulation, weather, and climatic trends. In turn, the consequences for humanity are potentially severe, yet under-estimated.

FORMATION AND ICE TYPES

There are many terms to express the different stages of sea-ice formation and dispersal. Many were introduced by mariners and indigenous people, and only a few of them are mentioned here. Sea-ice formation begins with the growth of millimetre-sized crystals, referred to as 'frazil' ice, as in rivers. These crystals coalesce into large masses as temperatures sink below the freezing point of sea water which, for average salinity, is -2°C (28°F).

The character of the ice, however, is dependent on the turbulence of the sea surface. In calm conditions, the crystals merge to form wafer-thin sheets of ice that have a dark greyish greasy appearance - hence the term 'grease ice'. The sheets readily slide over each other, creating thicker 'rafts'. We observe this process when a ship disturbs the newly formed ice, and it is fascinating to watch how sliding sheets inter-finger with each other as they are pushed together. As the ice thickens by this process, its appearance changes from black to grey; the thicker the ice, the lighter it appears. Up to ten centimetres (four inches) in thickness, this new ice is referred to as 'nilas'.

In choppy conditions, ice crystals merge into slushy pancakes. This distinctive 'pancake ice' grows into roughly circular rafts, typically a metre in diameter, and collisions between them result in upturned edges. Over time, nilas and pancake ice grow into more continuous and thicker sheets or floes, resulting in 'pack ice'. In the open sea, this ice is dynamic, but eventually can form a single sheet. Then, collision between large floes under the influence of wind and currents creates irregular 'hummocks' (and their underwater counterparts, 'bummocks'), and distinct 'pressure ridges'. The latter consist of broken blocks of ice, and typically range in thicknesses from 5 to 30 metres (16-100 ft), of which about a fifth is above water-level. Pressure ridges are a serious obstacle to surface travel, and, in the Arctic Ocean, many are the stories of explorers being delayed or defeated by such features.

Pressure ridges may be too thick for some icebreakers to break through; even a nuclear-powered icebreaker, the world's most powerful, may require several attempts to force a passage.

Whereas ice is compressed into ridges in some locations, in others the ice is stretched and breaks apart, forming linear openings called 'leads'. They range in width from a few metres to several kilometres. These, too, are a major obstacle to surface travel, but are a great advantage to ships looking for easier passage through the ice. Wider, more persistent openings are known as 'polynyas'. They form where there is upwelling of warmer water, or where offshore winds disperses the ice as soon as it forms. Leads and polynyas are a great attraction to marine mammals and bird-life.

Sea ice forms most readily in bays adjacent to land, to which it is anchored. The resulting sheet of ice is often smooth, and lacks pressure ridges. This is known as 'shore-fast ice'. Even before the sea freezes, ice builds up at the shoreline as sea spray from crashing waves freezes during autumnal storms. A small ice cliff, known as an 'ice foot' is created by this process.

Ice can also be classified according to age. 'New ice' embraces grease ice and nilas, 'first year ice' is thicker ice that has formed in the previous winter season, 'second year ice' is ice that has survived one summer melt season, and 'multi-year' ice is ice that has survived two or more summers. As ice becomes older, it undergoes an interesting

Types of young Arctic sea ice:
Young sea ice is referred to as grease ice, which has an oily appearance, and 'nilas', which is more brittle and initially has a dark appearance.
◀ *Nilas and grease ice forming Flyvefjord, a branch of Nordvestfjord in East Greenland (2017).*
▶ *Nilas and rafting of one plate over another observed through falling snow in the Arctic Ocean north of Franz Josef Land (2019).*
▶▶ *Stranded nilas in Nordbugta, a bay in Nordvestfjord, East Greenland (2017).*

Sea Ice 61

change. The salt, or brine, which occurs in new ice slowly drains out over time, so that in multi-year ice little is left. Mariners used this fact to obtain fresh water from pools on old ice. Most multi-year ice occurs in the heart of the Arctic Ocean, but the amount is diminishing rapidly. In Antarctica, multi-year ice is less able to form, because of the large reduction in sea-ice cover each summer under the influence of wind and currents.

SPRING-TIME SEA ICE BREAK-UP

During the course of the winter season, pack ice gradually thickens and becomes more extensive, although with interruptions when influenced by storms. In late winter as daylight increases and temperatures rise, sea ice begins to fracture, first forming leads, then splitting into separate angular floes. By spring, positive temperatures lead to surface melting and pools form, producing a turquoise/white mottled effect when observed in bright conditions from the air.

HISTORICAL OBSERVATIONS

Sea ice is a measure of the state of health of the planet, as it is one of the key indicators of climate change. Thus monitoring sea ice is vitally important. However, remoteness and the infrequency of visits to the polar regions meant that monitoring sea ice and establishing trends was unfeasible until the deployment of satellites in 1979. Until then, annual observations in the

▸ British Channel, Franz Josef Land, Russia, 2019
Angular ice floes observed in early summer during breakup of the winter sea ice cover. Ice-capped mountains are visible in the background.

▾ Near Iqaluit, Baffin Island, Canada, 2014
Early summer off the coast of southern Baffin Island, showing decaying ice floes with numerous small turquoise melt-pools.

○ **McMurdo Ice Shelf edge, Ross Sea, Antarctica, 2010**
Close to Scott Base, large pressure ridges form where sea ice impinges against the McMurdo Ice Shelf, producing a wonderland of ice pinnacles, ridges and snow drifts. A safe walking route through this area has been created for base staff to use.

Arctic were patchy, and in Antarctica almost non-existent. The earliest documented observations in the Arctic were made by the Vikings off the north coast of Iceland as early as 870 AD. From the mid-18th century whalers, sealers and explorers made sporadic observations. It was only in 1933 that the first systematic observations were made, by the Russians, and sea ice charts compiled. After the Second World War, shipping records and aerial photographic surveys have allowed several countries to compile ice charts. Most recently, the deployment of satellites has revolutionised our understanding of sea-ice extent and concentration, and in the last decade ice thickness as well. Both the Arctic and Antarctic are now well covered, as indeed are the other regions where sea ice forms.

EXTENT AND THICKNESS

Approximately twelve percent of the world's oceans are prone to sea-ice formation. In the Northern Hemisphere, sea ice forms mainly in the Arctic Ocean, which is centred on the North Pole. The Arctic Ocean is surrounded by land masses, and connects with the other world's oceans via the Bering Strait to the Pacific Ocean, and on either side of Greenland to the Atlantic Ocean. Historically, the Arctic Ocean has retained a substantial area of ice in summer, averaging about 7 million square kilometres (about 2.7 million square miles) by September, but effectively doubling to about 14-16 million square kilometres (about 5.4-6.2 million square miles) in late winter. However, in recent years these figures have been much reduced, as summarised below.

Beyond the Arctic Ocean, winter-only sea ice occurs in the Bering Sea and Sea of Okhotsk in the Pacific sector, Hudson Bay in northern Canada, the Greenland Sea, the Labrador Sea, and Baltic Sea in the Atlantic sector.

▲ Arctic Ocean, 2019
Where ice floes are forced together by wind and currents, the surface becomes hummocked and heavily fractured. With a polar bear for scale, this hummocked ice was observed in the Arctic Ocean north of Franz Josef Land, Russia.

In the Southern Hemisphere, sea-ice formation is largely limited to Antarctica. In contrast to the Arctic, Antarctica is a continent surrounded by the unconstrained Southern Ocean, defined by the cold waters that are formed around the continent. The Southern Ocean is linked physically to the Pacific, Atlantic, and Indian oceans, but is separated from them by a boundary between cold and warm water, known as the Antarctic Convergence. Sea ice grows extensively south of the Convergence in winter to cover an area averaging about 20 million square kilometres (about 7.7 million square miles), effectively doubling the ice-covered area of Antarctica each season. However, in summer it shrinks much more than in the Arctic, to just 4 million square kilometres (about 1.5 million square miles) on average.

SEA ICE TRENDS

To determine trends in sea ice extent, scientists tend to look at thirty-year averages, as variations between years are considerable. The baseline for the Arctic is the period 1981-2010, soon after satellite data became reliable. The trend for the minimum area of sea ice each season in September since then has been sharply downward, with the decline amounting to nearly thirteen per cent per decade. At the time of writing, the least sea ice coverage in the Arctic was in September 2012, when only 3.57 million square kilometres (1.38 million square miles) of sea ice survived, which was about half the long-term average. The figures for the decline in multi-year ice are even more striking. In 1985 nearly a fifth of the Arctic Ocean's sea ice was more than four years old, but by 2019 nearly all of this ice had gone. Thus renewal of Arctic ice has not been sustained. For these reasons, many scientists expect an ice-free Arctic Ocean to occur in summer within a few decades.

In Antarctica, the sea ice trends are unclear. Following a slow irregular increase since records began, around 1980, since 2017 there has been a sharp fall. There has been severe depletion in the West Antarctic sector, including the Antarctic Peninsula, where warming atmospheric and oceanic temperatures have been most notable. This more than offsets sea-ice expansion in other

◀ Arctic Ocean, 2019
Low summer sunlight emphasises the fractured nature of pressure ridges that are about 2m tall, as viewed from the deck of an icebreaker as it approaches the North Pole.

parts of Antarctica, where increased fresh water input (from warming) freezes more readily than undiluted sea water, or where cold offshore winds have increased.

Outside the polar regions, the most comprehensive sea ice records come from the Baltic Sea, which is surrounded by several countries: Sweden, Finland, Russia, Latvia, Estonia, and Poland. Because of the importance of shipping, data on sea-ice extent in the Baltic Sea can be traced back to 1720. Even today, icebreakers are needed to keep shipping lanes open in winter. The records show that the maximum extent of ice reached nearly 300,000 square kilometres (116,000 square miles) in the early 1800s, following which there has been an erratic decline, accelerating since the 1980s. 'Mild ice winters', defined as having less than 130,000 square kilometres (50,000 square miles) have become more frequent, with sixteen of them occurring in the period 1990-2019.

GLOBAL CLIMATE INFLUENCE

Sea ice has a major influence on the world's climate, and its fate should be a major concern to humanity. White, well-formed ice with snow cover reflects most of the sun's solar radiation back into space, in other words, it has a high 'albedo'. In contrast, ocean water is dark and absorbs much more solar radiation, thus warming the surface of our planet. As sea ice melts, pools develop on the ice surface, which also becomes darker than frozen ice. As the area of darker water grows in successive summers, the more the ice melts – this is known as a 'positive feedback effect'. Consequently, the ice in the Arctic Ocean is in what could be termed a death spiral. Within decades there may be no multi-year ice, or even second-year ice left in the Arctic Ocean, and in summer it may become largely ice-free. The surrounding lands are already becoming warmer at a rate that is several times more than the global average. Consequently, the glaciers and the Greenland Ice Sheet are melting at an accelerating rate (raising sea level), ocean currents are changing, and weather systems are being altered across the Northern Hemisphere.

Sea ice in Antarctica also has an important influence on oceanography. Along with ice shelves it introduces cold 'bottom water' into the global oceanic circulation. In turn, this influences global climate.

In conclusion, sea ice reduction is one of the major indicators of global heating, especially in the Northern Hemisphere. The consequences are the accompanying recession of glaciers and the Greenland Ice Sheet, deterioration of Arctic ecosystems, and changing Northern Hemisphere climate. Changes in the Antarctica are having a similar effect, at least in some areas, but several more years of data are needed to be sure of continent-wide trends.

▰ **Nordvestfjord, East Greenland, 2017**
Early autumn (September) in Nordbukta, a small bay in Nordvestfjord, where the initial freezing is forming 'nilas' near the shoreline and around the many icebergs that originate from outlet glaciers from the Greenland Ice Sheet.

◀ **Resolute, Cornwallis Island, Nunavut, Canada, 2022**
Fast ice stretches as far as the eye can see in this earl July view, looking southwest from a hilltop near Resolute Airport. Fast ice refers to winter sea ice that is connected to the land. It is commonly the last ice to melt in early summer, especially in bays, fjords, and between islands. At the fast ice edge numerous marine mammals and sea birds tend to congregate.

○ **Sedov Station, Franz Josef Land, Russia, 2019**
Fast ice with a slowly opening lead, adjacent to Sedov, the former large Soviet station in Tikhaya Bukhta on Hooker Island, one of the many islands in the Franz Josef Land Archipelago. Sedov was operational year round from 1929-1959. Today, only a few of the buildings are in use for ship-based tourism. The lead has formed in the decaying fast-ice during the early summer (June) melt.

◀ **Mt. Gaudry, Adelaide Island, Antarctica, 2012**
An ice foot near Rothera Station, a British base in Ryder Bay on Adelaide Island in the Antarctic Peninsula. In early winter, before the sea freezes over, spray from waves freezes on the adjacent shoreline. A substantial cliff of ice can remain well into the following summer season, as in this photograph taken on a fine December day. The heavily glacier-draped peak is Mt Gaudry (2531 m / 8304 ft).

▲ **Scott Base, Antarctica, 2010**
In the vicinity of New Zealand's research station, multi-year sea ice has been thrust upwards in a pressure ridge against the McMurdo Ice Shelf. The change in texture within the layers reflects successive freezing phases.

▶ **Helsinki, Finland, 2019**
Berthed at the port of Helsinki, a fleet of Finnish icebreakers are used to keep shipping lanes open in the Baltic Sea.

CHAPTER 6

GLACIER ICE

▶ Crusoe Glacier, Axel Heiberg Island, Nunavut, Canada, 2022
Glaciers come in all shapes and sizes, and there are many spectacular examples in the High Arctic. The tall cliff at the terminus of Crusoe Glacier indicates that the glacier was advancing until recently, but it has now begun to recede rapidly like all its other neighbours.

From the microscopic scale of exquisitely formed snowflakes, to the majestic glaciers that carve out mountain scenery, and to the massive ice sheets that drape most of Greenland and Antarctica, we are dealing with a single substance, the solid form of water, H_2O. What other substance displays such a wonderful variety of forms in Nature?

FROM SNOWFLAKES TO GLACIERS AND ICE SHEETS

It is sometimes hard to imagine that the delicate hexagonal symmetry of snowflakes falling from the sky provides the raw material for one of the most powerful forces of nature – glacier ice. Many parts of the world experience snowfalls that may last from hours to months, but only in high mountains and the polar regions does glacier ice form. It is in these regions that snow accumulates year-by-year without completely melting each summer.

Glaciers come in all shapes and sizes. The smallest include cirque glaciers and niche glaciers that are formed in mountain amphitheatres, and ice aprons and hanging glaciers that form on steep rock faces. On a larger scale there are long valley glaciers, highland icefields that spread unevenly between mountain peaks, and ice caps that form domes over high ground.

The largest ice masses are ice sheets, which by definition have an area of over 50,000 square kilometres (19,000 square miles). Today, the only ice sheets on Earth are those in Greenland and Antarctica. However, during the ice ages of the last two million years, large areas of North America, northern Europe and Siberia were also covered by ice sheets. Their legacy has strongly influenced the economic development of the affluent North by providing construction materials and a host for water supplies, as well as encouraging the growth of tourism in areas of glaciated landscapes (Chapter 14).

The flow of ice sheets is not uniform. Rather, there are certain narrow zones of fast-flowing ice called 'ice streams', that flow at speeds of several hundred metres a year. They are bordered by slow-moving ice, moving at only a few metres a year. In fact, it is ice streams that discharge most of the ice into the surrounding ocean. In addition, there are slabs of land-derived ice floating on the ocean called 'ice shelves', the largest in Antarctica being the size of France. In the last few decades many ice shelves have collapsed as the ocean and atmosphere has warmed, producing huge tabular (flat-topped) icebergs that are sometimes the size of small countries.

TRANSFORMATION OF SNOW TO GLACIER ICE

Glacier ice forms where seasonal snow survives through successive summers and accumulates year-by-year. The amount of snow that accumulates varies widely - from a few centimetres a year in the middle of the Antarctic Ice Sheet to several metres a year in high-alpine environments. As layers of snow become buried by new snowfalls, the snowflakes are transformed into denser crystals with a sugary texture, called 'firn'. With deeper burial, firn crystals, in turn, recrystallise into glacier ice crystals. Glacier ice is typically full of air bubbles and in temperate regions the crystals can grow to several centimetres in length. The annual layers of snow turned into ice can be preserved for long distances downglacier, and may even survive as far as the terminus, especially around the edges of ice caps where ice flow is not constrained by valley sides.

Scientists who have studied the chemical composition of air bubbles in ice cores drilled right through the polar ice sheets have found that the trapped air represents the atmosphere at the time of burial. By measuring greenhouse gases in each layer of ice in turn, the scientists have obtained a remarkable record of climate change spanning the last 800,000 years. This is how we know, for example, that carbon dioxide levels are now 40% higher than at any time over this period, as a consequence of burning of fossil fuels and other human impacts since the onset of the industrial revolution.

▸ **Greenland Ice Sheet, 2022**
The vastness of the Ice Sheet, also known as the Inlandsis (Danish), is apparent in this aerial photograph of an outlet glacier, Eqip Sermia, in West Greenland. Numerous crevasses indicate the fast-flowing and dynamic nature of the ice sheet at this location.

▲ Northwest Greenland, 2012
Shrinking ice caps are increasingly common in the High-Arctic, such as this small one to the northwest of the main Greenland Ice Sheet. The contrast between the white snow-covered accumulation area and the greyish ice of the ablation area is clearly evident. Many of these accumulation areas are now disappearing as the climate heats up.

Glacier ice has a density that is nine-tenths of that of water, and commonly has a pale blue appearance. In contrast, ice that is derived from water, such as within crevasses or at the glacier bed, commonly has few bubbles and appears dark blue. The blue colour derives from the fact that ice absorbs all colours of the light spectrum, except blue, which is reflected. Air bubbles lessen, but do not remove totally this effect. Thus, in ice cliffs or in icebergs we may see striking blue veins of bubble-free ice contrasting with the delicate pale blue of bubble-rich ice.

GLACIERS AND CLIMATE

Glaciers form in a wide range of climatic conditions. In the interior of Antarctica or in arid coastal areas, where snow accumulation is just a few centimetres a year, temperatures are so low that melting is limited. Thus, the ice remains at a temperature well below the melting point. Unsurprisingly, such glaciers are referred to as 'cold'. In many parts of Antarctica, net snow accumulation even occurs at sea level, and mass-loss is achieved by iceberg-calving rather than melting.

○ Gornergletscher, Valais, Switzerland, 2018
The view from the tourist viewpoint at Gormergrat near Zermatt. Gornergletscher is one of the largest valley glaciers in the Alps, and is surrounding by many famous mountains, including Monte Rosa (4634m / 15,200 ft) on the left, Lyskamm 4533 m / 14,870ft) and Breithorn (4164 m / 13,660 ft) in the centre, and Matterhorn (4478 m / 14,690 ft) at far right. Medial moraines separate the different flow components of this composite valley glacier, and the glaciers on Breithorn (centre) are now separated from the main glacier.

At the other extreme, glaciers in the European Alps, the Southern Alps of New Zealand, the Western Cordillera of North America, Iceland, and Scandinavia, a winter snowfall of several metres is reduced by melting and rapid increase in density, and most of the glacier ice remains at a temperature close to the melting point. Glaciologists refer to these glaciers as 'warm' or 'temperate'. In these mountain regions long, steep glaciers often extend down into the forest zone. The association of blue glacier ice and green forests, backed by snowy peaks is a magnificent sight. An in-between temperature state prevails in the Arctic and on the highest

Our Frozen Planet

▲ **George VI Ice Shelf, Antarctic Peninsula, 2012**
Ice shelves are commonly featureless white plains of snow-covered floating ice. However, in this view the ice shelf provides added interest in being sandwiched between the Antarctic Peninsula mainland (background) and Alexander Island (foreground).

○ **Hardinger Icefield, Alaska, 2012**
An example of a highland icefield on the Seward Peninsula, southern Alaska. Such ice masses differ from ice caps in having an undulating surface with mountains projecting above them. The mountains that are surrounded by ice are known as 'nunataks'.

mountains on Earth, notably in the Himalaya and Andes. In these regions, glaciers have a mixture of 'cold' and 'warm' parts and are referred to as 'polythermal' glaciers.

Temperature differences are reflected in the physical appearance of glaciers and the nature of meltwater runoff, as well as in the landforms they produce. Meltwater from cold glaciers has limited impact on the landscape, and the glaciers themselves are not very effective erosional agents, even though they can deform frozen ground. Arctic polythermal glaciers tend to carry relatively little surface debris, but a large amount is carried at the base and deposited directly as 'till'. Melt-streams are powerful, but usually confined to the sides of the glacier, where they may cut deep gorges. Temperate glaciers, in contrast, carry a large amount of surface debris, while most meltwater is concentrated at the glacier in a central tunnel which emerges at the 'snout' or 'toe'. Most glacial sediment is then reworked by meltwater, the resulting rivers producing wide braided river plains where channels constantly split and re-join, changing rapidly during the course of a summer season.

By looking at glaciers, we can readily visualise the effect of climate change. The position of the snout or toe advances or recedes in response to climate change averaged over the previous several years or decades. This behaviour masks year-by-year variability of climate. One of the primary tasks for the glaciologist is to ascertain the state of health of a glacier, or its 'mass balance'. The principle is that snow builds up in the accumulation area at high altitude, converts to ice, which then flows under gravity into an ablation area, where there is a net loss of ice through melting. A positive mass balance, measured over a year, is when accumulation exceeds net melting or 'ablation'. A series of positive mass balance years will translate into an advance of the glacier snout, and we can regard such a glacier

Glacier Ice

as being 'healthy'. Conversely, a negative mass balance is when ablation exceeds accumulation, and a series of such years will lead to thinning and recession of the snout. Today, most glaciers have a negative mass balance, and indeed many are about to vanish as a consequence of global heating.

GLACIER FLOW AND DEFORMATION

Glaciers flow down-slope under the influence of gravity. The upper layers of a glacier are brittle, and this attribute is reflected in a variety of fractures. Crevasses form where ice is under tension, such as where a step in the glacier bed produces an icefall. In temperate glaciers crevasses may open to depths of 30 metres (100 ft), but considerably more in cold glaciers. Crevasses are death-traps to the unwary, but climbers are generally trained in roping up and extricating each other if they fall in. Crevasses are often hidden by snow bridges, so traversing snow-covered glaciers demands that the correct precautions are taken. Intersecting crevasses give rise to ice towers called 'séracs', and such areas are best avoided as they are prone to unpredictable collapse. Many people have died through falling into crevasses, or from collapse of séracs.

In contrast, at depth and under pressure, ice behaves approximately like a material that can bend, which in physical terms approximates to 'plastic' deformation. We can imagine how ice in a glacier flows in the following experiment: allow fairly thick porridge to flow in a confined channel down a slope! This experiment demonstrates how movement is faster in the middle than at the sides, and faster at the top than at the bottom. Plastic deformation produces a variety of structures in glaciers at depth. As the glacier surface melts in the ablation area, these deep structures are beautifully exposed, giving each glacier a unique character. A layered structure called 'foliation' is the most evident of these structures. Foliation on smooth, rain-washed ice has the appearance of polished marble, or when weathered by exposure to sun it is reminiscent of a ploughed field. Folds are another feature of plastic deformation, and these can range in size from less than a metre to hundreds of metres (a few to several hundred feet), forming elegant sweeping arcs across the ice surface. Geologists recognise that these structures resemble those found in rocks that were deformed deep down in the Earth's crust and have since risen to the surface.

Some structures are formed by a complex mix of brittle and plastic deformational processes. Spectacular curving bands of light and dark ice called 'ogives' are an example of this. Ogives typically form below icefalls, where they consist of foliation, folds and fractures. Pairs of dark and light layers are often formed annually, and are the product of both internal deformation and fracturing.

In temperate regions, a glacier not only deforms internally, but also slides over its bed. This 'basal sliding' is facilitated by meltwater at the glacier bed. The water arrives there from the surface via channels and potholes or "moulins". In some temperate glaciers in the Alps, measurements have shown that basal sliding accounts for as much as 80% of the total movement. Naturally, because more meltwater is present, most of this movement takes place in summer. By contrast, sliding in polythermal glaciers is much less, whereas cold glaciers are mostly frozen to the bed and most movement is by internal deformation.

IMPORTANCE OF GLACIERS

Glaciers are the life-blood of many dry regions of the world, as they provide a reliable water resource for irrigation and hydro-power generation (Chapter 13). They also provide landscapes that are much-loved by tourists, and can host areas for summer skiing. For both economic and aesthetic reasons it is vital that we conserve what we can of the world's glaciers.

◀ Southern Axel Heiberg Island, Nunavut, Canada, 1977
Outlet glaciers from icefields on southern Axel Heiberg Island in the Canadian High-Arctic, flow through narrow bedrock channels and spread out in the lowlands as piedmont glaciers.

▝ **Steel Glacier and Mt Steele, Yukon Territory, Canada, 2006**
Steep-sided mountains in areas of high snowfall generate unstable masses of ice referred to as ice aprons. This image shows Mt Steele (5073 m / 16,644 ft), the fifth highest mountain in North America. Heavily crevassed ice aprons descend the steep north face to join the valley glacier below. Major rockfalls have been recorded from this peak, and the image shows evidence of some small-scale events.

◀ **Aoraki / Mt Cook, Southern Alps, New Zealand, 2008**
Hanging glaciers, such as these below the summit of New Zealand's highest mountain (3724 m / 12,218 ft), cling to precipitous mountainsides, occasionally releasing unpredictable ice avalanches.

▲ **Barnes Ice Cap, Baffin Island, Nunavut, Canada, 2022**
Glaciers are generated through the year-on-year accumulation of layers of snow referred to as stratification, which, as a consequence of burial, is transformed through firn into glacier ice. If internal deformation is limited, and ice is transferred passively towards the terminus, stratification can survive right through the ice mass. This high-altitude aerial image of the edge of the Barnes Ice Cap, shows annual layering remarkably well preserved. The darker layers have been influenced by melting and by concentrating of dust.

▶ **Breiðamerkurjökull, Iceland, 2015**
Through the process of burial and compaction, recrystallisation, and flow, glacier ice crystals can grow into irregular shapes several centimetres across. Here, large interlocking crystals are exposed near the snout of Breiðamerkurjökull, a large outlet glacier from Iceland's largest ice cap, Vatnajökull.

Glacier Ice

▲ **Glacier du Chardonnet, Argentiére, Mont Blanc Massif, France, 2008**
Small glaciers carve out mountain amphitheatres known as cirques. This small cirque glacier is situated in the catchment of Glacier d'Argentiére in Western Europe's highest mountain massif.

◂ **Lake Mathieson, Southern Alps, New Zealand, 2008**
From mountain top to forest: glaciers descend from Mt Tasman and Aoraki/Mt Cook in New Zealand's Southern Alps, viewed from Lake Mathiesen. In many mountain regions, the contrast between the blue shades of glaciers and the green of the surrounding forests is striking.

Glacier Ice

◯ **Paradise Bay, Antarctic Peninsula, 2019**
In many parts of Antarctica, there is net accumulation of snow and ice down to sea level, and most mass-loss of ice is through calving into the sea, as here in Paradise Bay on the west side of the Antarctic Peninsula.

▸ **Palmer Land, Antarctic Peninsula, 2012**
Flow in glaciers is both brittle and plastic, near the surface and at depth respectively. Within the uppermost steep parts of a glacier, the main mass of ice can separate from ice clinging to a rock face, forming a deep crevasse referred to as a bergschrund. The bergschrund in this image is visible below the fluted snow and rock outcrops.

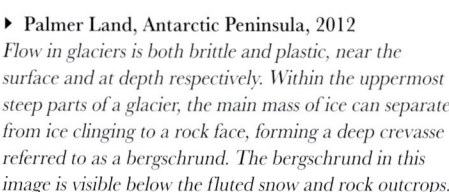

◯ **Fox Glacier, Southern Alps, New Zealand, 2008**
Brittle deformation in the form of crevasses is most evident in fast-flowing ice masses, including those that are topographically constrained, as evident in this aerial view of the upper part of Fox Glacier.

Glacier Ice 93

◀ **Austre Brøggerbreen, NW Spitsbergen, Svalbard, 2013**
Evidence of plastic deformation in glaciers in once deeply buried ice is in the form of a layered structure called foliation. Running parallel to the ice-flow direction, this longitudinal foliation in a small valley glacier was formed at depth as ice was squeezed into a narrow channel.

▼ **Alpefjord, Northeast Greenland National Park, 2018**
The plastic nature of ice at depth is well illustrated by the folding of the debris layers in the combined terminal cliff of Gully Gletscher (left) and Søfstrøm Gletscher (right) as they calve into the fjord. The folding was inherited from deep ice higher up the glacier, but the ice at the terminus is now affected by brittle deformation in the form of narrow crevasses.

○ **Mer de Glace, Mont Blanc Massif, Chamonix, France, 2018**
A striking feature of some glaciers with icefalls is a set of curving light and dark structures called 'ogives'. They are the result of strong compression at the base of an icefall, and their curved form reflects an important aspect of glacier flow - that ice flows fastest in the middle. The Mer de Glace is the largest glacier in the French Alps, and shows a fine set of ogives. These structures were the subject of investigation by some of the earliest glaciologists in the mid-19th century.

CHAPTER 7

DISAPPEARING GLACIERS & SHRINKING ICE SHEETS

◀ Vadret del Forno (Forno Glacier), Canton Graubünden, Switzerland, 2011
A small valley glacier, set amongst the granite peaks of the Bregalia Range. The smooth, gentle slope of the snout is characteristic of a rapidly receding alpine glacier.

Most mountain glaciers around the world, the Greenland Ice Sheet, and parts of the Antarctic Ice Sheet are all shrinking rapidly in response to global heating. The decline of mountain glaciers is evident from the retreat of their snouts and overall thinning. Indeed, it is projected that areas such as the Alps will lose almost all of their glaciers by 2100. The shrinking of ice sheets is evident from around forty years of satellite measurements.

RECOGNISING GLACIER ADVANCES AND RECESSIONS

How do we know mountain glaciers are receding? Firstly, we can tell from their visual appearance; secondly by direct measurement of mass balance, which is the year-on-year balance between accumulating snow and melting as described in Chapter 6; and thirdly from aerial photography and satellite images.

Glacial geologists are the scientists who assess the visual appearance of glaciers, and examine the landforms surrounding them. From a range of attributes, they can 'read' how the glacier is behaving, and can even interpret landscapes in areas where glaciers have long-since disappeared.

An advancing, 'healthy' glacier shows the following characteristics: a steep bulging snout, evidence of active flow such as crevasses, a convex transverse surface profile, and an end-of-summer snowline at a relatively low level on the glacier. Conversely, a glacier that is in 'poor health' or in recession shows a gently graded and a smooth or rotting snout, has few crevasses and a concave transverse surface profile. The unhealthy glacier has a snowline that recedes high up the glacier by late summer, or even disappears altogether, demonstrating the loss of its accumulation area. Also, large areas are riddled with meltwater channels and subglacial cavities, progressively eating away at the glacier's heart.

Other evidence of recession, in the immediate surroundings of a glacier, include trim-lines, which mark the former level a glacier reached, lateral moraines along the valley sides comprising unstable glacial debris, and recessional moraines which are piles of debris left behind at the snout of the glacier (Chapter 9). There are also ice-cored moraines where stagnant or dead ice is covered by loose debris, and extensive areas of bare ground that have not yet had time to be colonised by plants.

RECORDING GLACIER CHANGES

The longest records of glacier change come from the European Alps, where paintings record the state of glaciers near to the Little Ice Age peak, around 1750 to 1850. Examples are the works of Zumbühl and Fearnley depicting the glaciers near Grindelwald. Snout positions of around a hundred glaciers in Switzerland have been measured on the ground annually by farmers, teachers, historians and clergymen, and all these data have been assembled by the Swiss Glacier Monitoring Service. Some records even extend back 150 years. The data show that, although there has been a serious decline in all glaciers over this period, some of them advanced for short periods, the most recent being in the mid-1970s to early 1990s.

Photographic documentation is vitally important in determining the amount of recession or advance, and provides us with among the most powerful visual impressions of climate change. Using images repeated from exactly the same place several years or decades apart, combined with historical photographs and even paintings, we are able to demonstrate how glaciers have changed over time.

To quantify the state of health of a glacier, or its 'mass balance' (Chapter 6), glaciologists measure input as snow, and output as ice-melt. Satellite data nowadays have largely superseded the traditional method of inserting stakes into the ice to measure surface-level changes. Around the world, data from representative or 'index' glaciers, with measurements going back more than thirty years, have been assembled. They provide a global overview of the state of health of the world's long-lived ice masses on land. An organisation called the World Glacier Monitoring Service (WGMS), based in Zürich, Switzerland, operating under the auspices of several international bodies, collects and analyses the data annually.

▸ **Unteraargletscher, Berner Oberland, Switzerland, 2009**
A large glacier complex with a tongue that is deeply buried in debris, marked by the curving stripes of different coloured rocks. The glacier is undergoing severe recession, marked by the stagnating, pitted snout. Thinning of the glacier since the Little Ice Age of the early 19th century is indicated by the pale grey 'tidemark' or 'trimline' on the mountain flanks. The lake in the foreground has been dammed for hydro-electric power generation. The peak in the background is Lauteraarhorn (4042 m / 13,261 ft), with its steep glaciers no longer connected to the main glacier.

Mountain glaciers are particularly useful indicators of climate change because they fluctuate in terms of length and thickness according to long-term trends (measured over decades), rather than year-to-year variations, Thus, the trends recorded by WGMS are of global significance. Since 1950 there have been only five years when the combined records from these index glaciers have had a positive mass balance, and none since 1987. Furthermore, the general trend since then has been increasingly negative. For example, the tongues of large Alpine glaciers have typically receded several kilometres and lost more than 200 metres (656 ft) of thickness since 1850, with most of this happening in the last few decades.

A visit to any mountain region which still has glaciers reveals the tell-tale signs of recession. Glaciers in Alaska, western Canada, and the USA, and the European Alps show particularly steep declines, and their attractiveness to visitors is declining as the concentrations of surface rubble increases. An estimated ninety percent of the world's glaciers are receding. Glacier advances are few in number, with an anomalous cluster is evident in the Karakorum Range of south-central Asia. Here, the reason is that glacier mass balance is influenced not only by temperature, but also by precipitation. Increased snowfall in glacier accumulation areas may be brought about by warmer, moister conditions, and this may translate into an advance of the snout.

Globally, a small percentage of glaciers behave in a curious way, inasmuch that they periodically undergo a rapid advance or 'surge', lasting from a few months to a few years. Concentrations of surge-prone glaciers are found in in Alaska, the Yukon, and Svalbard. These glaciers respond to changing dynamics and conditions at the bed, rather than to climate. For instance, changes in the distribution of temperature and meltwater at the bed affect how the glacier flows. The surge involves the rapid transfer of ice from the accumulation area to the ablation area, and is not associated with a gain in mass. Between surges the glacier simply stagnates *in situ*.

Despite the small number of exceptions, the available data clearly indicates that rising global temperatures are having a profoundly negative effect on mountain glaciers worldwide. The implications for sea-level rise, water resources, irrigation, and hydro-electric power generation, not to mention tourism, are deeply worrying.

▲ Crusoe Glacier, Axel Heiberg Island, Nunavut, Canada, 2008
Advancing glaciers are increasingly few and far between. This glacier was still advancing in 2008, but when visited in 2022 it was already receding.

▲ South Croker Bay Glacier, Devon Island, Nunavut, Canada, 2022
The Devon Island Ice Cap generates several glaciers that reach the coast along Lancaster Sound, the eastern part of the route of the famed North West Passage. This tidewater glacier, like its neighbours, is receding rapidly.

CHANGES TO THE GREAT ICE SHEETS

Of course, it is not the demise of mountain glaciers alone that is having an impact globally, with their potential to raise sea level about half a metre by 2100. More concerning is what is happening to the great ice sheets of Greenland and Antarctica. We have few historical records regarding ice-sheet growth and decay, and it is only since the late 1970s that satellite data have begun to fill the gaps in our knowledge. For a long time it was thought that ice sheets responded only slowly to climate change, but that view is changing on account of startling evidence emerging from both those ice sheets.

The Greenland Ice Sheet in recent years has shown accelerating ice movement, increased discharge into the ocean as icebergs, widening of the area of ablation, and rapid recession of its outlet glaciers that terminate in the sea (Chapter 10). Overall, scientists have shown that ice was being lost from Greenland seven times faster in 2020 than it was in the 1990s. The unprecedently hot summer in Greenland in 2019 triggered the loss of 600 billion tonnes of ice from the ice sheet in just two months, which was enough to raise sea level globally by an average of 2.2 millimetres. Events such as these cast doubt on prior estimates of sea-level rise of up to one metre by the end of the century, estimates which will probably need revising substantially upwards. Melting of the Greenland Ice Sheet, and to a lesser extent the other Arctic glaciers, is also having a profound effect on circulation in the North Atlantic. The addition of more fresh water into the ocean is slowing down the Gulf Stream, which moderates the climate of northwest Europe. A weakened Gulf Stream could, ironically, see colder winters affecting this region, even though the rest of the world is heating up.

The Antarctic Ice Sheet shows more complex behaviour. Indeed, scientists divide it up into three unequal ice sheets, the East, West, and Antarctic Peninsula ice sheets, each behaving differently. The East Antarctic Ice Sheet is by far the largest mass of ice, but is the least known. Radar data indicates that it largely rests on land above sea level. Therefore, glaciologists have tended to consider it to be fairly stable. Much of it is fringed by ice shelves, which grow out into the ocean slowly, but then break off in large chunks, before resuming outward growth. Thus, patterns of advance and recession are difficult to assess on the human time-frame. Some satellite data tends to show possible thickening of ice in the interior, from increased snowfall, but net loss around the fringes. There is also increasing awareness that some zones

▲ Baby Glacier, Axel Heiberg Island, Nunavut, Canada, 2008
Many glaciers around the world are losing their accumulation areas and are thus not being sustained. The absence of a snowfield in the upper reaches of this small glacier is a good indication that the glacier is 'dying', and like many glaciers is likely to disappear in the coming decades.

▲ Aktineq and Fountain glaciers, Bylot Island, Nunavut, Canada, 2014
Ice aprons with no accumulation areas, lying between these two large glaciers, are all that remains of a once-healthy glacier on the near lefthand mountainside. A frozen ice-dammed lake is situated between the two main glaciers.

Disappearing Glaciers & Shrinking Ice Sheets

of fast-flowing ice, known as 'ice streams', are showing signs of destabilisation (thinning and accelerating) as they are attacked from below by encroaching sea water from the warming Southern Ocean.

Glaciologists are much more concerned about the fate of the West Antarctica Ice Sheet, which has been viewed as a candidate for 'collapse', since it is largely grounded below sea level, and is being eaten into by the warming ocean. Major floating ice streams, flowing into the Pacific segment of the Antarctic Ocean, have shown dramatic signs of thinning and recession in recent years. A large joint UK-US programme started in 2018/2019, studying Thwaites Glacier, which is the size of Great Britain or Florida. It is revealing very rapid recession as a consequence of ingress of warm water underneath. From satellite data, the amount of ice flowing out of this region into the ocean has nearly doubled over thirty years. Thwaites Glacier on its own has the potential to raise sea level over half a metre. If it disappears, and the rest of the ice sheet destabilises, it could cause a further two or three metres of sea level rise.

The Antarctic Peninsula Ice Sheet, although only a fraction of the size of the other two, has shown the most dramatic changes in recent decades. Ice shelves have disintegrated one after the other, mainly on the west and northeast sides. Since they already float on water, ice shelves do not add to sea-level rise in themselves. However, when they disintegrate their buttressing, or holding-back, effect on land-based ice in the interior is removed. It is the resultant increase in discharge of land-based ice into the ocean that adds to sea-level rise. Collapsing ice shelves have been linked to the several degrees of measured climatic warming in the Antarctic Peninsula since the 1950s. However, since the late 1990s temperature has decreased somewhat, a change that is linked to extreme natural variability of atmospheric circulation. Nevertheless, a recent study of glaciers in the region has shown that 95% of glaciers are receding, reflecting the longer-term temperature trends.

PAST AND FUTURE TRENDS

Overall, then, as temperatures have risen globally, so have glaciers and ice sheets receded. The trend has been most obvious in mountain regions such as the Alps, where the volume of ice loss has been a quarter to a third of their volume in the last few decades. The visual quality of the Alps and other mountain regions has already been negatively affected, as glaciers are more strewn with rubble. By the end of the 21st century, when most of the glaciers have gone, we will be left with large areas of vegetation-poor, rubble-dominated valley floors, although reforestation will ultimately heal and create a very different landscape. The acceleration of ice loss since the 1990s shows no signs of abating. The ice sheets, equally, are suffering massive losses, way beyond what was predicted just a few years ago. The consequences are already being felt globally, as sea levels rise.

▸ **Vadret da Tschierva (Tschierva Glacier), Pontresina, Switzerland, 2008**
This glacier shows dramatic evidence of recession, especially in the 21st century. It has receded back towards the source mountain group of Piz Bernina (4049 m / 13,284 ft), revealing large lateral moraines that were formed during the Little Ice Age of the 18th and early 19th. The pine trees were growing on the terminal moraine produced by the glacier at that time. However, on 14 April 2024, a massive rockfall, with an estimated million cubic metres of debris, fell from a steep rock face in the middle background. Debris mixed with snow and ice obliterated all the vegetation in the foreground. The run-out distance of this huge landslide was 5 km (3 miles), but fortunately there were no human casualties.

▸ **Khumbu Glacier and Pumori, Nepal, 2003**
In the Everest region of Nepal, the Khumbu Glacier, like many in the Himalaya, is debris-mantled. The thick layer of debris is slowing the thinning of the glacier, but not preventing it. Ponds and lakes are increasingly developing on such glaciers, and these have the potential to become unstable and cause outburst floods.

◯ **Grosser Aletschgletscher, Berner Alpen, Switzerland, 2018**
The largest glacier in the Alps, looks at its best when surrounded by the colours of autumn. A trimline on the far side of the glacier, where pale rock and thin vegetation contrast with darker more weathered rock and richer vegetation, marks the ice level during the early 19th century, late in the Little Ice Age.

◯ **Vadret da Morteratsch (Morteratsch Glacier), Engadin, Switzerland, 1985-2021**
Progressive recession of the snout of Vadret da Morteratsch, an accessible glacier descending from the peak of Piz Bernina. The glacier was photographed in 1985 (top), 2007 (middle) and 2021 from the exact same location. Within this time period the glacier receded 1.2 kilometres, and the bare ground left behind is being colonised by carpets of flowering plants and species of tree (including willow, larch and pine). Alongside a popular walking route to the location of the glacier snout in 2010, dated marker posts mark the recession since 1850. The ground beyond to the current terminus (2023) has only an intermittent rough bouldery alpine trail.

◂◂ **Kongsbreen, NW Spitsbergen, Svalbard, 1996 & 2013**
These two photographs, taken from exactly the same position on Ossian Sarsfjellet in 1996 and 2013 respectively, show how the glacier Kongsbreen has receded. The expansion of a bay with a calving glacier front is evident in 2013, while the ice surface beyond has dramatically thinned.

▾ **Comfortlessbreen, Spitsbergen, Svalbard, 2010**
A small percentage of glaciers around the world, notably in the High-Arctic, Alaska-Yukon and the Karakorum, undergo sudden, unpredictable phases of advance, known as surges, which are related to dynamic instability, rather than climate change. Svalbard has the highest concentration of surge-type glaciers. In this view, the glacier Comfortlessbreen is coming to the end of a 5-year surge, and demonstrates the characteristic steep active terminus, ploughing into the ground in front.

▼ Wahlenbergbreen, Isfjorden, Spitsbergen, Svalbard, 2018
The large valley glacier Wahlenbergbreen was surging into Yoldiabukta, a branch of Isfjorden, as recently as 2022, after several years of advance. This 2018 photograph shows the steep terminal face and the heavily crevassed nature of the glacier surface.

Changes at the glacier base, from frozen to wet, are thought to be one of the triggers for a surge in Svalbard, especially when a large volume of meltwater reaches the bed and lubricates it. The expansion of the glacier tongue is accompanied by thinning in the accumulation area, so these glaciers are not actually 'growing' in volume.

▲ **Fjørtende Julibreen, NW Spitsbergen, Svalbard, 2022**
Svalbard has a remarkable number of surge-type of surge-type glaciers, with a handful surging at any one time. However, 2021 and 2022 saw a dramatic increase in the number of surging glaciers to more than 20, one example of a new surge being that of Fjørtende Julibreen (Fourteenth of July Glacier).

◯ **Daugaard-Jensen Gletscher, Nordvestfjord, Northeast Greenland National Park, 2017**
The Greenland Ice Sheet is increasingly responding to climate change through recession of its outlet glaciers and thinning of its surface. Daugaard-Jensen Gletscher is one of the largest outlet glaciers on the east coast of Greenland, and is notable for delivering huge icebergs into the world's longest fjord, Nordvestfjord.

▲ **Smeerenburgbreen, NW Spitsbergen, Svalbard, 2022**
This large tidewater glacier originates from a highland icefield and is fed en route to the sea by several smaller glaciers. Its name means "Blubber town glacier", after the site of a former Dutch whaling station. The prominent peak on the left, Hornemantoppen (1073 m; 3520 ft) is a 'horn', formed as a consequence of glacial erosion on all sides.

▼ Paradise Harbour, Gerlache Strait, Antarctic Peninsula, 2018
Recent studies of the Antarctic Peninsula using satellite imagery demonstrate that 95% of the region's glaciers are receding. This tidewater glacier is responding to changes in accumulation on the Antarctic Peninsula Ice Sheet, the smallest of the three ice sheets in Antarctica.

Disappearing Glaciers & Shrinking Ice Sheets

CHAPTER 8

MELTING GLACIERS

▶ **Fountain Glacier, Bylot Island, Nunavut, Canada, 2014**
Unusually heavy summer rainfall resulted in significant changes to the runoff characteristics of the glacier. Water reaching the bed, mixing with basal sediment, backed up, and under high pressure escaped through various orifices at the glacier surface. Heavy rainfall events are becoming more common in the High Arctic.

All glaciers undergo melting to some degree, with meltwater providing the glacier-traveller with the opportunity to observe an array of fascinating phenomena. Meltwater also plays an important role in changing the landscape around a glacier, and can contribute to the economic well-being of an area. Most critical of all is that glaciers are one of the best indicators of climate change, as they respond to warming or cooling temperatures on a time scale of decades, rather than annually. As glaciers are currently losing mass, their current contribution to sea-level rise is already having an impact on humanity around the world.

In winter there is little melting, as sub-zero temperatures predominate, and the sounds of wind and blizzard prevail. In contrast, summer is a time when snow and ice melt, and then the background noise we experience in the vicinity of a glacier is the sound of running water: tinkling, gurgling, rushing, and roaring. The sound of falling rocks is also common, as frost-shattered rocks, held together by ice, disintegrate as the thaw begins. Glaciers everywhere generate their own stream systems, both on their surface, or within and below the ice. In fact, glacier stream systems resemble the three-dimensional networks found in limestone regions, with numerous englacial tunnels. Visitors to a glacier will often face a variety of obstacles and hazards to getting around, including streams flowing off the ice, as well as deep channels and potholes on the glacier surface itself.

CONTROLS ON MELTING

The principal influence on glacier melting is, naturally, air temperature. However, even on bright sunny days with sub-zero temperatures, melting may be significant because of intense solar radiation, even in the frigid polar desert regions of Antarctica. Most glacier-covered regions of the world experience daytime summer temperatures several degrees above freezing. During fine weather positive temperatures and the sun's radiation generate large volumes of meltwater, but at night there is usually an almost total freeze-up. Consequently, discharge of water from a glacier fluctuates daily. A practical aspect of these fluctuations is that, whereas a meltwater stream may be easily crossed in the morning, by mid-afternoon it may have become impassable. In cloudy weather daily extremes are less pronounced and melting at night may be only slightly less than during daytime. Meltwater production also varies on a seasonal basis and is usually greatest in late spring and early summer. Even in the sub-zero conditions of winter, run-off may continue because of the release of internally stored water.

Another agent promoting melting is geothermal heat - the heat from Earth's interior. It normally only produces limited melting at the bed of a glacier but is still enough to influence the flow-dynamics of the ice by causing the ice to slide on its bed more easily. Exceptionally, volcanic regions provide extreme examples of geothermal heating. Large volumes of ice can melt rapidly, particularly during subglacial volcanic eruptions, and create subglacial lakes, which may burst in spectacular fashion. This process happens every few years in Iceland. Frictional heat, generated as a glacier slides over its bed, or as a result of internal deformation of ice crystals gliding over one another, can also generate meltwater.

Melting on and around a glacier usually begins in spring, although rainfall events in winter are increasingly common as the climate heats up. Melting may not be immediately visible because water collects at the base of the snowpack. Eventually, the snow becomes saturated if the water cannot drain freely, such as on flat surfaces, and a snow swamp may form. Such areas are dangerous and best avoided, but they may be difficult to see with the naked eye.

GLACIER DRAINAGE

Once the snow has melted from the glacier surface, a system of veins develops at the boundaries between ice crystals. This allows the meltwater to percolate through glacier ice, and even develop a 'water table' below the surface. However, most water migrates through a glacier by way of a network of much bigger channels and conduits, although some collects on the surface as ponds. The development of meltwater channels on the surface of a glacier depends on various factors: the rate of melting, the rate of deformation and temperature of the ice, the extent of crevasses and the pattern of other structures such as foliation (the layered structure in ice), and ice temperature. Surface channel

▶ **George VI Ice Shelf, Antarctic Peninsula, 2012**
Surface melt on Antarctic glaciers commonly takes place in sub-zero temperatures when solar radiation is absorbed by debris, forming ponds. Refreezing in shaded areas can produce beautiful icicles as here at the edge of George VI Ice Shelf adjacent to Ablation Lake on Alexander Island.

○ *Thompson Glacier, Axel Heiberg Island, Nunavut, Canada, 2022*
Aerial view of supraglacial meltwater ponds (black with cryoconite sediments) and turquoise surface streams, reflecting clean ice underneath. The main channel is slightly incised into the near flat glacier surface, and is approximately 2 meters (6ft) wide.

systems develop best on stagnant and on cold glaciers, whereas on heavily crevassed glaciers surface drainage is swallowed up by these open fractures. The channels themselves range in size from tiny rills to canyons several metres deep and a hundred or so metres wide. On flat, crevasse-free glaciers, the streams may form into a dendritic pattern – resembling the branches and trunk of a tree. Alternatively, streams may form tight, meandering, deeply incised channels with overhanging walls and meander cut-offs. These are one of the two principal obstacles to glacier walking, the other being crevasses. Once the water reaches the bed, it may form a large tunnel, perhaps even the size of an underground railway tunnel.

Glacier drainage is commonly influenced by the distribution of surface debris and ice structures such as foliation. Medial moraines are lines of debris that form when two glaciers join. They often stand out as ridges because the debris protects the ice beneath from melting, and streams are thus constrained to the edges of the moraine. When moraines provide a source of debris, sediment may be transported substantial distances in low-friction channels. Streams may also flow parallel to foliation, since different ice types melt at differing rates, forming a characteristic ridge-and-furrow topography. Crevasses and other fractures provide routes for meltwater to penetrate deep into the glacier, and they are sites where a glacier mill or 'moulin', similar to a pothole in limestone country, can form. Much of a glacier's surface meltwater reaches the bed *via* moulins. Standing on the edge of a moulin is a disconcerting experience. It is as though the falling water is drawing you closer to the slippery edge to see where it is going.

GLACIER WATER-STORAGE AND RELEASE

Small pools of standing water may develop on the flatter reaches of a glacier, especially where dark patches of debris or dust absorb more solar radiation and melt down into the ice. The smallest examples of pools are cylindrical tubes a few centimetres across, and sometimes up to tens of centimetres deep; they are known as 'cryoconite holes'. The sediment plays host to bacteria, and these tiny pools become hot-spots of microbiological activity. The holes merge to form larger ponds, and, in extreme cases, lakes tens of metres or yards across. Large lakes, kilometres in length, seem to develop almost exclusively on glaciers in the polar regions. Notably, those on the Greenland Ice Sheet are implicated in accelerating melting since the darker water absorbs more solar radiation than the surrounding ice. The meltwater also speeds-up the outlet glaciers as the water drains to the bed *via* moulins.

Apart from collecting on the surface, glaciers store water in a variety of situations in the form of ice-dammed, proglacial, and subglacial lakes. Ice-dammed lakes occur where a stream from a side valley meets the glacier, or at the confluence of two glaciers. Ice-dammed lakes commonly fill up during the melt-season until water pressure is so great that the glacier is raised from its bed, allowing the water to escape beneath the ice, causing a flood.

Proglacial lakes form in low-relief areas that are in contact with the ice, or behind moraine dams in high mountain regions. Although the former are usually benign, the latter in regions such as the Andes and Himalaya have the potential to cause major catastrophes if the dam is breached and collapses (Chapter 13).

Other large ice-contact lakes are those that form subglacially as a result of volcanic activity, as in Iceland. When subglacial volcanoes erupt large amounts of meltwater are produced. The resulting flood is referred to as a "jökulhlaup", and many such events have been documented from southern Iceland over several centuries. Such floods occur on an almost unimaginable scale, and they produce large, constantly changing flood plains called "sandar" ("sandur" singular) from the Icelandic language. The development of settlements and permanent infrastructure on sandar is too risky, although in Iceland the country's ring-road could only be completed by

◄ *Morteratsch valley, Engadin, Switzerland, 2018*
During periods of melting, especially on sunny days, streams leaving a glacier show strong variations in discharge, as this pair of images taken below the glacier Vadret da Morteratsch demonstrates. Early morning shows a relatively low discharge and sediment load, while by mid-afternoon the opposite is true.

Melting Glaciers

○ **Crusoe Glacier, Axel Heiberg Island, Nunavut, Canada. July 2022**
The impressive terminal cliff of a glacier that has recently begun to recede. On a warm summer's day, large volumes of sediment-laden meltwater flow across the front and sides of the glacier, undercutting the ice cliff, causing it to collapse, while at the same time causing considerable erosion of the adjacent hillsides. Note the person for scale on the left.

building sections across these plains. Road sections and bridges are unavoidably destroyed by major jökulhlaups but are usually rebuilt rapidly afterwards.

Meltwater is an important component of the glacial environment. It is responsible for much of the erosion that occurs beneath and at the margins of a glacier, and also produces a wide range of depositional landforms. Such features are not only an important component of the glacial landscape, but meltwater deposits are valuable as sand and gravel resources.

IMPACT OF GLACIER MELT ON HUMANITY

Meltwater from glaciers is currently a vital resource for billions of people on our planet. Glaciers in high-mountain regions such as the Himalaya, Andes, Alps, Southern Alps of New Zealand, and Western Cordillera of North America deliver water, into the rivers that flow into the lowlands, for drinking, irrigation, and the generation of hydro-electric power.

Himalayan glaciers act as a 'water tower', and provide seasonal meltwater to over a billion people, *via* several of the major rivers in Asia, including the Indus, Ganges, Brahmaputra, Yangste, Yellow, Irrawaddy and Mekong. Glaciers of the Andes provide meltwater for the capital cities of Lima in Peru and La Paz in Bolivia. On the negative side, meltwater stored within unstable glacial lakes has proved responsible for devastating floods that have costs the lives of thousands of people.

In the Alps and in Norway, glacier meltwater is stored in reservoirs and harnessed for the generation of hydro-electricity through a complex network of tunnels. The fate of these mountain glaciers should be of deep concern to humanity. It is a resource that is declining rapidly - the 'water towers' are not being fully replenished each winter season.

Perhaps the greatest impact of glacial meltwater is from its increased production as climate heats up. As glaciers and ice sheets lose mass, the resulting meltwater is the major contributor to global sea-level rise, and the rate is accelerating, resulting in increased vulnerability to flooding for millions of people living in low-lying coastal communities.

◀ **Fountain Glacier Bylot Island, Nunavut, Canada, 2014**
Deep incision by a supraglacial (surface) meltstream has produced a canyon of impressive proportions.

▶ **Austre Brøggerbreen, NW Spitsbergen, Svalbard, 2013**
Supraglacial streams are commonly clean, but if they flow over debris, the fine sediment carried in suspension can colour the water according to the source rock. This image of merging supraglacial streams reflects water-flow over red sandstone (left) and greyish-green metamorphic rock (right).

◀ **Fox Glacier, Southern Alps, New Zealand, 2011**
Surface meltwater on the temperate Fox Glacier is finding its way to the base via a vertical shaft called a 'moulin'. These impressive chasms form where there is a structural weakness in the ice, such as a former crevasse.

▶ **Breiðamerkurjökull, Southeast Iceland, 2015**
A glacier guide inside a subglacial meltwater channel beneath the outlet glacier from Iceland's largest ice cap, Vatnajökull. The water level is low in spring, enabling ready access for a few hundred metres from the glacier margin. Meltwater in temperate glaciers tends to migrate to the base rather than the sides.

○ Vadret da Morteratsch (Morteratsch Glacier), Engadin, Switzerland, 2009
Meltwater has carved an englacial tunnel into the glacier. In winter, a descent through the tunnel became possible, as no meltwater was then flowing.

▲ Oberaargletscher, Berner Oberland, Switzerland, 2023
A glacier portal at the snout of Oberaargletscher. This large subglacial tunnel, carrying most of the meltwater from the glacier, is the source of the River Aare, a major north-flowing river in Switzerland.

▲ Rob Roy Glacier, Southern Alps, New Zealand, 2011
The temperate Rob Roy Glacier clings to a rockface above the Matukituki Valley. When subject to strong melting the outflow feeds several prominent waterfalls, commonly reaching peak-flow in mid-afternoon.

▶ Thompson Glacier, Axel Heiberg Island, Nunavut, Canada, 2008
The terminal cliff of Thompson Glacier displaying a major englacial outflow of sediment-rich meltwater. The brown water shows that there is a hydrological connection to the glacier bed. The water emerges along a plane of weakness, such as a fault, within the ice. The lower dark grey zone in the cliff is debris-rich basal ice.

▲ **Eqip Sermia, Greenland Ice Sheet (West), 2022**
An aerial view of a meltwater lake near the edge of the Greenland Ice Sheet at the outlet glacier known as Eqip Sermia. Several meltwater streams flow towards the lake, which is estimated to be about 100 metres in length, while other streams flow in different directions. The black material is dust and algae, and pin-points a closed moulin, through which the water will eventually drain. The increasing number of surface lakes and streams on the Greenland Ice Sheet has a profound effect on glacier flow when they drain to the bed.

▶ **Gornergletscher, Valais, Switzerland, 2011**
Small thin patches of debris on the surface of a glacier absorb more solar radiation than the reflective ice around, creating 'cryoconite holes'. Those on Gornergletscher near Zermatt, are exceptionally well-developed for the Alps. The sediment in the bottom of the hole, 'cryoconite', commonly has a diverse microbiota, including bacteria that clumps the sediment grains together.

◯ **Thompson Glacier, Axel Heiberg Island, Nunavut, Canada, 2022**
Highly colourful rocks surround tributary glaciers to the large Thompson Glacier (foreground right). Ice-dammed lakes full of icebergs are slowly forming as the tributaries recede. The lakes in this image are Five-Finger Lake (in the foreground) and Phantom Lake (background right).

Melting Glaciers

CHAPTER 9

NATURE'S DEBRIS CONVEYOR

◂ **Tokositna Glacier, Alaska Range, Alaska, 2008**
A fine set of medial moraines on Tokositna Glacier, descending from Denali (6190 m / 20,310 ft), the highest mountain in North America. Debris from rockfall is sourced along the valley sides which, where two tributary glaciers combine, forms a medial moraine. Glacier flow transfers a large amount of debris towards the snout. Multiple moraines are the result of several tributaries joining the main glacier.

Glacier ice is an enigmatic material that on the one hand is so vulnerable to melting and disappearance, yet on the other can literally 'move mountains'. In fact, there are few natural processes on Earth's surface that can match a glacier's capability in transporting debris a long distance away from its source.

DISTRIBUTION OF DEBRIS IN GLACIERS

One of the most obvious features of a typical mountain glacier is the amount of rubble that litters its surface. This rubble, consisting of angular rocks of all sizes, comes from rockfall from the surrounding peaks and from the action of the glacier scraping along the valley sides. A glacier that is healthy tends to have many crevasses which ingest much of the surface debris. Conversely, an unhealthy, shrinking glacier sees the gradual concentration of debris at the surface, until it becomes totally covered, and according to some may look no more attractive than a gravel quarry. Material that is carried at the surface of a glacier is referred to as 'supraglacial debris', but even more debris may be carried out of sight, at the glacier base. This basal debris, embedded in the ice, is the reason that glaciers are such powerful erosive agents. Thus, a typical mountain glacier is capable of carrying a huge volume of debris, and it may be considered as a massive conveyor belt, transporting material from all points along its length towards the snout. Ice sheets and ice caps, although having few mountains projecting above them, also carry large volumes of debris, but this is carried out of sight, almost entirely in the basal zone.

DEBRIS TRANSPORTED AT THE SURFACE OF A GLACIER

Each type of glacial debris has distinctive attributes, and together they contribute to the landscape character on and around glaciers and ice sheets. Supraglacial debris is the product of rockfall derived from exposed rock along the flanks and at the head of a glacier, where freeze-thaw processes fracture the rock. Much of this debris is buried in the accumulation area, where there is a net gain in mass from snowfall. Then, when the debris re-emerges in the ablation zone, where there is a net loss of ice from melting, it occurs in many different forms – as isolated boulders, as boulders on ice pedestals known as glacier tables, as piles of sediment in the form of dirt cones, and as ridges. Debris commonly completely covers the glacier surface, which if thick enough can slow down ablation. Although most rockfalls are small, occasionally entire landslides cover broad expanses of ice, especially in regions prone to earthquakes, such as the Western Cordillera of North America, the Andes, the Himalaya, or the Southern Alps of New Zealand.

Most debris falling down mountainsides becomes caught up in the ice as the glacier slides past the valley sides. The resulting lines of debris at the edge of the glacier are known as 'lateral moraines'. Where two glaciers join, the individual lateral moraines combine to form a 'medial moraine', and the line of debris extends towards the snout in mid-glacier. Where several glacier tributaries join, multiple medial moraines form. In most cases, these medial moraines remain parallel to the glacier sides, but if glaciers are prone to short-lived erratic advances or surges (Chapter 6), the moraines are contorted.

Walking across the surface of a debris-strewn glacier yields fascinating insights into glacier behaviour. The debris itself is blocky or angular in shape, and during summer when the ice is melting it is unstable. Large blocks can suddenly slide or rotate without warning, so walking across such a surface requires care. A cover of debris can slow down the rate of ablation compared with clean ice, so a medial moraine often forms a ridge standing proud several metres above the glacier surface.

The area of debris in relation to exposed ice increases towards the snout on most mountain glaciers. Uneven melting of ice beneath the debris cover and the action of streams or subglacial rivers on slow-moving or stagnant ice create an irregular surface of sharply defined hills and hollows containing pools, with a relief of several metres. We call this 'dead ice topography' or 'glacier karst'.

▲ Pasterzenkees, Hohe Tauern National Park, Austria, 2009
A glacier table, where a rock sits on an ice pedestal, has formed on Pasterzenkees, the largest glacier in the Austrian Alps. While the surrounding ice melts, the boulder greatly reduces solar radiation beneath and, thus, protects the ice from melting.

▶ Unteraargletscher, Berner Oberland, Switzerland, 2005
Debris (or dirt) cones are generated where supraglacial streams have collected sediment and sorted it. The finer-grained components of sand and fine gravel collect in hollows, which then protects the ice from ablation, ultimately resulting in an upstanding drape of sediment over the ice.

Nature's Debris Conveyor   147

◤ **Kongsbreen and Ossian Sarsfjellet, NW Spitsbergen, Svalbard, 2009**
A prominent curving ridge, made up of basal glacial sediment and stagnant ice, has formed where Kongsbreen impinges against Ossian Sarsfjellet, Svalbard. The ridge dates from around 1900 AD and holds back a moraine-dammed lake.

◣ **Nabesna Glacier, Wrangell Mountains, Alaska, 2012**
Converging tributary glaciers carry debris from rocky slopes in the foreground, forming medial moraines as the valley glacier flows towards the lowland in the distance.

◀ **Kennicot Glacier, Wrangell Mountains, Alaska, 2012**
Irregular hummocks of debris-covered dead glacier ice, referred to as 'glacier karst', and scattered ponds at the snout of this receding glacier.

▶ **Vadrec del Forno, Maloja, Switzerland, 2011**
Glaciers in mountain regions transport large volumes of rockfall debris on their surface. The fragments are angular and may reach several metres in size. These large boulders, resting on the glacier surface, are the result of rockfall from the surrounding granite peaks.

DEBRIS TRANSPORTED AT THE BASE OF A GLACIER AND INTERNALLY

Debris carried at the base or 'bed' of a glacier is very different in character from supraglacial debris, since it is continually being modified as the ice moves downhill. Freeze-thaw processes at the bed fracture the bedrock, and the products are readily picked up by the glacier. The assimilated debris forms the 'basal ice layer', which is able to abrade the solid rock beneath, with the help of sliding induced by meltwater. As the stones rotate in the ice, their edges are broken off and rounded, and along with the bedrock they are scratched ('striated') and polished, with some being pulverised into powder, or 'rock flour'. The resulting sediment is a random mixture of particles of all sizes from clay or silt up to boulders several metres across. After deposition it forms a layer over the land called 'basal till'. The finer particles are washed out in subglacial streams, often giving the meltwater a distinctive milky appearance. If sediment is flushed directly into the sea or a lake, then plumes of sediment with distinct boundaries with the surrounding clear water occur.

In addition to the supraglacial and basal debris, most glaciers contain englacial debris, which originates from both the surface and base of the glacier. Supraglacial debris can be ingested when crevasses open up, or when streams wash it down moulins. Debris is also recycled from the glacier bed by deformational processes in the ice, such as by thrusting and folding of the basal debris layer.

CLIMATE CONTROLS ON DEBRIS DISTRIBUTION

The proportions of debris carried at the glacier surface or bed vary according to the temperature distribution in the glacier. Temperate glaciers, that is those at their melting point throughout and typically found in alpine regions, carry most of their load at the surface. In contrast, polythermal glaciers, which are those with both ice below and at the melting point, and are found mainly in the polar regions, carry the bulk of debris at the base. Cold glaciers, which have ice below the melting point throughout, are frozen to the bed, yet ice-flow can still deform the debris underneath.

OTHER SOURCES OF DEBRIS

Debris in glaciers also includes dust that has been carried by the wind. Sometimes the dust is whipped up by storms in far-distant lands, and carried aloft in the upper atmosphere. For example, yellow Saharan dust is blown onto glaciers in the Alps several times a year, staining the glaciers orange. In other regions, notably Iceland and Alaska, volcanic eruptions have thrown substantial amounts of ash on to a glacier. If the time of an eruption can be determined, the ash layer can tell glaciologists a lot about glacier mass balance and dynamics.

Other types of debris and chemicals in glacier ice are, by volume, insignificant, but are important in providing clues about environmental change, such as traces of industrial pollution or forest fires (soot), and radioactive fall-out from nuclear explosions. In addition, microbiological processes result in the growth of bacteria and algae at the surface, staining the ice a variety of colours, especially grey, red and green. All these enhance the rate of melting of the ice, as the darker colours absorb more solar radiation than clean ice.

Evidence for the processes outlined above is retained in the landscape when the glaciers have receded. Lateral and medial moraines are the most prominent features in the early stages of recession as they maintain a core of stagnant glacier ice, commonly separated from the main glacier. Over time, as the ice melts, they become more subdued, but are still recognisable unless destroyed by slope processes such as landslips (Chapter 11).

In summary, the multiplicity of processes involved in debris transport and deposition by glaciers means that our 'debris conveyor' is constantly sorting, mixing and recycling sediment as it moves.

▲ Kongsbreen, NW Spitsbergen, Svalbard, 2009
The orange-brown staining in this image of Kongsfjorden is the result of subglacial discharge from the glacier, Kongsbreen, in the background. The sediment is suspended in the water and commonly forms a distinct plume that stretches out from the glacier margin.

▶ **Fountain Glacier, Bylot Island, Nunavut, Canada, 2014**
Glaciers are noted for recycling debris as they flow over the land. Here, a scientist at the margin of Fountain Glacier is intrigued by the complex processes whereby basal debris, indicated by the dark layer with a large boulder, is elevated above the bed.

◐ **Conwaybreen and Kongsbreen, NW Spitsbergen, Svalbard, 2009**
View across debris-covered Kongsbreen towards Conwaybreen Kongsbreen is a rapidly receding tidewater glacier, that has left stagnant ice along the flanks of the fjord in which it terminates. This stagnant ice is rich in debris that was originally uplifted from the glacier bed to the surface, as indicated by the rounded nature of the boulders and extensive areas of mud.

▼ Glacier d'Argentière, Mt. Blanc massif, France, 2008
Saharan dust stains the firn in the accumulation area of this glacier. Dark layers of earlier dust events are preserved within the firn stratigraphy visible on the edges of large blocks of firn, known as sèracs.

◀ Buckskin Glacier, Alaska Range, Alaska, 2012
Debris from two landslides on Buckskin Glacier, is slowing down ablation. Since the ice around the debris is melting much faster, the landslides appear to rest on an elevated platform.

▶ Tasman Glacier, Southern Alps, New Zealand, 2008
An aerial view of the debris-mantled surface of the country's largest glacier. A large rockfall in 1991 that lowered the summit of the nation's highest peak, Aoraki/Mt Cook by 10 metres (33 ft), descended the tributary glacier (lower right) and created a lobe of boulders that stretched across the full 2 km width of the glacier. It was estimated that 12-14 million cubic metres (16-18 million cubic yards) of snow, ice and rock had been involved in the avalanche.

Nature's Debris Conveyor

CHAPTER 10

WHERE GLACIERS MEET THE SEA

▶ Magga Dan Gletscher, Gåsefjord, East Greenland, 2017
Many of the outlet glaciers of the Greenland Ice Sheet are receding, and some are emerging onto land at the head of the fjords, as in the case of Magga Dan Gletscher. The glacier now partly terminates on a rock cliff and releases spectacular falls of ice into the fjord below.

The interaction between glaciers and ice sheets, and the ocean, is one of the defining characteristics of the polar regions. High mountain ranges, which border the coast in more temperate regions, such as in southeast Alaska and Chilean Patagonia, also generate glaciers that terminate in the sea. The loss of ice from these marine-terminating or 'tidewater glaciers', a process we call calving, produces magnificent ice cliffs, backed by ice full of crevasses.

Calving glaciers are a major draw for tourists, and the number of cruise ships visiting to see such sights has been on an upward trend for many years. When the media seeks to illustrate 'global warming', they often choose to illustrate the calving process, but this process occurs whether or not the glacier shown is advancing or receding, and the relationship with climate is more complex than for land-based glaciers.

ICEBERGS FROM ANTARCTIC ICE SHELVES

The largest glaciers that terminate in the sea are ice shelves – thick floating slabs of land-derived ice that may reach several hundred metres in thickness. They are mostly confined to Antarctica, but a few survivors of the last Ice Age are found in the Canadian Arctic and North Greenland. Antarctic ice shelves commonly show horizontal layering in their cliffs. This is an important feature of many Antarctic glaciers, which is that they receive snow accumulation right down to sea level.

Many ice shelves that fringe the Antarctic Ice Sheet do not grow and recede in response to cooling and warming respectively. Rather, they grow outwards slowly and then, maybe after several decades, a large chunk breaks off as a tabular (flat-topped) iceberg. Nowadays, scientists track large icebergs (known as mega-bergs) by satellite and give them numbers. The largest iceberg ever recorded was B-15, which broke off the Ross Ice Shelf in 2000. It measured approximately 295 km in length and 37 km in width, having a surface area of 11,007 km^2, which is roughly the size of Jamaica. This iceberg broke into large pieces, and by 2006 several approached the coast of New Zealand, while another one had reached beyond South Georgia as late as 2018. As B-15 demonstrated, Antarctic icebergs have a long life-span, often lasting a decade or more, when they are confined to the cold Southern Ocean. However, when they escape northwards beyond the cold/warm water oceanographic boundary known as the Antarctic Convergence, they melt rapidly. The drift of large tabular icebergs has occasionally encouraged business people and arid nations to develop proposals to drag them to water-stressed areas of the world, but so far to no avail.

The last half century has seen warming of both atmosphere and ocean around the Antarctic Peninsula, and this has led to the loss of several major ice shelves in the region, a process that continues to this day. Attacked from underneath by warm ocean currents, and from above by meltwater, these ice shelves have collapsed in dramatic fashion, producing 'armadas' of icebergs.

ICEBERGS FROM FLOATING ARCTIC GLACIERS

In the northern hemisphere, large glaciers that float when they reach the sea are constrained between valley walls, such as are found in deep fjords, notably in Greenland. They are a type of tidewater glacier, but like ice shelves they also produce tabular icebergs. The largest-recorded iceberg of this type was a slab which formed in 2010 after breaking off Petermann Glacier in northwest Greenland, and then drifted south towards Newfoundland. It covered an area of approximately 260 km^2, roughly the size of the English City of Birmingham. The sinking of the *Titanic* in 1912, with the loss of approximately 1500 lives, was a

◤ **Northwest Weddell Sea, Antarctica, 2012**
The Antarctic Ice Sheet, and especially the ice shelves that fringe it, produces the world's largest icebergs. As the ice is already floating when calving occurs, the icebergs are tabular in form and can be many kilometres long. This pair of tabular bergs off the coast of the Antarctic Peninsula in the northwest Weddell Sea, are just one or two kilometres in length.

▸ **Meares Glacier, Prince William Sound, southeast Alaska, 2012**
Terminating in the sea, this glacier is almost constantly calving and producing small icebergs and bergy bits, with bigger events occurring unpredictably with less frequency. The glacier also reveals how it ploughed into mature forest at its true right margin, prior to receding, and revealed bedrock stripped of vegetation.

▲ Hansbreen, Hornsund, southwest Spitsbergen, Svalbard, 2018
Like the termini of most tidewater glaciers in Svalbard, Hansbreen rests directly on the sea-bed, and the icebergs that are produced tend to be small and rarely larger than a bus. This early summer view shows extensive snow cover remaining on the glacier and the surrounding mountains.

consequence of a collision with one of these icebergs. Many of these glaciers are fast-flowing and originate in the heart of the Greenland Ice Sheet. Consequently, they are heavily crevassed, and the resulting tabular icebergs, which are commonly a few kilometres long, have multiple pillars (séracs) and sides that look as if they have been chiselled.

CALVING ICE WALLS AND TIDEWATER GLACIERS

The term 'ice wall' is used in a rather specific sense by glaciologists, for ice cliffs that are sitting directly on the sea bed, often in relatively shallow water. The ice behind the wall is generally slow-moving, and calving does not produced tabular bergs, but rather small irregularly shaped icebergs and small fragments at infrequent intervals. Ice walls are common all around Antarctica, and those in the Antarctic Peninsula will be especially familiar to expedition ship passengers.

○ **Nordvestfjord, Northeast Greenland National Park, 2018**
In the northern hemisphere, tabular icebergs are only formed in the deepest fjords where the glacier tongues that produce them float. Nordvestfjord has among the largest and most beautiful icebergs in the Arctic, as this image of a heavily crevassed berg illustrates. They show a wide range of blue tones, which are especially prominent in cloudy conditions. The iceberg preserves the intensely crevassed surface of the glacier, from which it previously calved.

◂ **Nordvestfjord, Northeast Greenland National Park, 2018**
A smaller iceberg, floating in calm water, allows one to see its underwater extension, including the projecting spur known as a keel. Icebergs such as this are unstable, and with seven eighths of the mass below water, they can flip over suddenly – an event that needs to be avoided when travelling in small boats.

◂ **Nordvestfjord, Northeast Greenland National Park, 2018**
The disintegration of a large iceberg is illustrated by this image. Crevasses are forming to the sides of the arch, while slabs of ice are peeling away from the arch's roof. Collapse seemed imminent but failed to materialise while we were on location.

In the Arctic the longest stretch of ice wall is in Nordaustlandet, Svalbard, which stretches for 180 km (112 miles). Sporadically fast-flowing (surging) ice streams from the interior ice cap breach the wall occasionally and generate local concentrations of icebergs.

The most dramatic glaciers with terminal ice cliffs that rest directly on the sea floor are valley glaciers that emanate from high interior mountains. Also referred to as tidewater glaciers, they have numerous crevasses and calve along these ready-made fractures. A calving event creates spectacular splashes and tsunami-like waves that are a big draw for tourist vessels. It is not advisable to camp on shorelines close to such glaciers! Southeastern Alaska is famous for tidewater glaciers, and there are many more in the Canadian and Eurasian Arctic islands, in Patagonia, and in the Antarctic Peninsula. The icebergs produced in these regions by the calving processes of peeling off and forward-toppling from these grounded glacier margins are relatively small, generally less than a hundred metres (300 ft) long. Usually, there is little warning of a calving event, so people approaching them by boat need to give such glaciers a wide berth.

ICEBERG CHARACTERISTICS

Icebergs come in all shapes and sizes, and are produced by several different processes. They are uniquely beautiful, ranging from white to pale blue to deep blue or turquoise. Some even are black from debris or algae.

Icebergs derived from floating tidewater glaciers are sublime architectural wonders, forming castellated walls of ice, tables, spires, towers (even twin towers), caves, arches and mushroom shapes. The largest may stand fifty metres (160 ft) or more above the waterline, commonly reaching higher than the top deck of a passing ship. Considering that seven-eighths of a berg is below water, those sailing out of the deepest fjords may exceed 500 metres (1600 ft) in total thickness. Arguably, the most beautiful of these tabular icebergs are found in Nordvestfjord in East Greenland, where the fast-flowing outlet glacier, Daugaard-Jensen Gletscher, terminates as a floating tongue in waters exceeding 1000 m (3000 ft) deep. These icebergs rapidly disintegrate as they drift slowly out of the fjord, producing spires, towers and arches of stunning blue and turquoise colours.

Icebergs, of course, are vulnerable to attack from wind, currents, waves, melting and refreezing. They also calve themselves. Their centre of gravity may change as a result, leading to tilted waterlines or complete turning over - adding to the subtle changes in texture of the ice.

Smaller fragments of ice are also associated with large icebergs, and old nautical terms have been used to describe them. 'Bergy bits' are small icebergs several metres (a dozen feet) across, while brash ice is a cluster of fragments less than a metre (3 ft) or so in diameter. Bergy bits that are made up of bubble-free ice or contain debris, and wallow low in the water, are referred to as 'growlers' – from the noise they used to make when struck by a wooden vessel in days gone by.

THE SCIENTIFIC CHALLENGE

Glaciologists find that trying to understand the relationship between marine-terminating glaciers, iceberg production, the ocean, and climate is a major challenge. What is not in doubt is that many such glaciers are in recession and producing more icebergs than in earlier decades, whether these be ice shelves in the Antarctic Peninsula, ice streams in West Antarctica, or tidewater glaciers in Alaska and the Arctic.

◀ **Bjorne Oer, Scoresby Sund, East Greenland, 2017**
Icebergs that are the shape of a tower and a pyramid have become grounded. As they disintegrate, bergy bits and brash ice covers the sea surface.

◀ **Nordvestfjord, Northeast Greenland National Park, 2018**
Icebergs are prone to rotation and reveal lines that mark undercutting at the waterline. This iceberg in Nordvestfjord, East Greenland has rotated clockwise, revealing one prominent and one weakly developed waterline. The column-like structures perpendicular to the upper waterline are bubble rills, which are formed by the release of air bubble from the glacier ice below the waterline.

▶ **Nordvestfjord, Northeast Greenland National Park, 2017**
With a backdrop of some of the world's oldest rocks, 2-3 billion year old gneisses, this pinnacled iceberg is reflected in the calm waters of the deep fjord.

○ **Hodge Glacier, Northwest Greenland, 2022**
An aerial view of the terminal region of Hodge Glacier, an outlet from the Greenland Ice Sheet. Although the ice sheet is shrinking, the glacier is highly dynamic, as illustrated by extensive crevassing. The dark double stripe is a medial moraine, formed where two separate ice streams join. In the lower part of the image, there are fewer crevasses and melt-streams are flowing towards the terminus. Considerable volumes of meltwater reach the bed of the glacier and discharge into the sea via subglacial tunnels, the outflows from which are indicated by the pale brown and orange-brown sediment plumes in the upper and lower part of the image respectively.

▲ **Melchior Islands, northern Antarctic Peninsula, 2019**
The seas surrounding the Antarctic Peninsula have many relatively small icebergs that are supplied by glaciers descending from the Peninsula Ice Sheet as well as from the mountains on adjacent islands. Alternating positive and sub-zero temperatures, even in summer, result in the formation of beautiful icicles, as on this iceberg.

▼ **Paradise Harbour, Gerlache Strait, Antarctic Peninsula, 2019**
The variety of iceberg shapes is infinite and even small ones like this can have strange shapes. This mushroom-like berg shows an uplifted waterline, below which erosion by the sea has been more pronounced than that exposed to the atmosphere for longer.

▲ **Paradise Harbour, Gerlache Strait, Antarctic Peninsula, 2019**
Much of the Antarctic coastline is fringed by 'ice walls', which are formed where ice on land flows slowly into a shallow sea, commonly where the underlying bedrock is exposed at low tide. Calving from such cliffs is rarely dramatic, as generally only small pieces of ice peel away at a time. In this late evening view is of ice walls near Paradise Bay, sunlight is striking the top of the cliffs with an orange glow.

▶ **Kronebreen, Kongsfjorden, northwest Spitsbergen, Svalbard, 2013**
Calving events from tidewater glaciers are unpredictable, and unless one is lucky, much waiting and a close watch is needed, especially when the event has already happened when the sound reaches the observer. In this case we were able to take multiple exposures of a calving event from a n active, fast-flowing glacier in Kongsfjorden, showing the sometimes explosive nature of this process.

▼ Conwaybreen, northwest Spitsbergen, Svalbard, 2009
As icebergs disintegrate into smaller fragments known as bergy bits and brash ice, tidal processes and wind can concentrate them on beaches, as here in the vicinity of the calving face of Conwaybreen.

▼ Kongsfjorden, northwest Spitsbergen, Svalbard, 2013
Icebergs melt and weather in ways that reflect their internal structure. This iceberg, stranded on the shore of Kongsfjorden near the science township of Ny-Ålesund, shows a furrowed surface that is the result of ice layers in foliation weathering at different rates.

○ Lilliehöökbreen, Krossfjorden, northwest Spitsbergen, Svalbard, 2018
The magnificent calving terminus of Lilliehöökbreen stretches across a large bay at the head of Krossfjorden, where it delivers a large volume of icebergs, bergy bits and brash ice into the sea.

CHAPTER 11

THE LEGACY OF GLACIERS & ICE SHEETS

◀ **Matterhorn from Theodulgletschersee, Valais, Switzerland, 2018**
The world-famous Matterhorn, reflected in a lake once close to Theodulgletscher, near Zermatt. The mountain lies on the Swiss-Italian border and epitomises extreme glacial erosion. Steep ridges, including the Hörnli Arete on the right, rise up to the pointed summit. The 4,478 m (14,692 ft) high 'horn' is one of the highest peaks in the Alps, and has been subject to glacial erosion on all four sides. As climate heats up, the ground ice which binds the rock together is melting, triggering rockfalls, and creating dangerous conditions for climbers.

Glaciers are noted for creating some of the most dramatic and beautiful landscapes on Earth. In many of our mountain regions, the effects of glacial erosion are well-displayed, even though the glaciers have long since disappeared. Sharp peaks, narrow ridges and steep-sided, flat-bottomed valleys and fjords, are among the many illustrations of the effectiveness of glacial erosion. These mountain regions also possess a variety of piles of sand, gravel and mixed sediments, which are the product of glacial deposition. Even more abundant glacial deposits are found in adjacent lowland regions, or on the plains that underlaid the last great ice sheets, such as in North America, northern Europe and Siberia.

GLACIAL EROSION

The effects of erosion can be seen on all scales - from small outcrops of bedrock to the world's highest peaks, and to vast areas of ice-scoured low rocky country. The distinctive imprint left by glaciers allows us to identify those regions that were covered by ice thousands or even millions of years ago.

The largest erosional features are spectacular glacial troughs, which include flat-bottomed, steep-sided valleys with lakes and marine inlets or fjords. They form the core of landscapes that mountain-lovers enjoy, especially in the more accessible mid- to high-latitude mountain regions, such as the Alpine countries, Norway, the British Isles, western North America, New Zealand, and Patagonia. Above the valley floors we may encounter upland amphitheatres, carved out by ice and known as 'cirques' (or 'corries' in the UK). Higher still are 'horns' and 'arêtes' which, respectively, are glacier-carved peaks that are pyramidal or horn-like in shape and narrow ridges. Differences in erosion between a main valley glacier and a tributary give rise to a 'hanging valley', often linked by an impressive waterfall. Low-lying areas of ancient resistant bedrock, with a landscape of low hills and small lakes, is referred to as 'knock-and-lochan topography' and is the product of 'areal scouring' by ice sheets.

Meltwater erosion commonly accompanies erosion by debris-laden glacier ice. Considerable volumes of water are generated in summer from melting snow and ice. In temperate glaciers this water reaches the bed *via* moulins and tunnels, and here it is under so much pressure that it is highly erosive, forming subglacial gorges over time. In polar regions, meltwater is forced to the edges of the glacier, and here the streams erode 'ice-marginal channels'.

At an intermediate scale of tens to hundreds of metres, the subtle processes show how the tools carried by a glacier – the boulders and finer debris embedded in the ice – have been at work. 'Roches moutonnées' are bedrock mounds, a few tens of metres high and long, with a smooth striated up-valley-facing slope, and a rough jagged down-valley-facing slope, which are the product of abrasion and bedrock fracture respectively. If meltwater under high pressure at the base of a glacier combines with direct glacial erosion, a range of grooves, hollows and exceptionally deep potholes forms in the bedrock.

On bedrock surfaces at the smallest scale, we can observe a range of distinctive features, including 'striations' (scratch marks from debris embedded in the ice) and a wide range of other erosional markings. Representatives of all these features are preserved in the geological record, and allow us to reconstruct the dimensions and flow patterns of former ice masses.

GLACIAL DEPOSITION

The products of glacial deposition complement the features of glacial erosion, and commonly occur side-by-side in highland areas. Although less dramatic than erosional landforms, they are nevertheless distinctive components of the landscape. In lowland areas, depositional landforms give rise to the rolling countryside with fertile fields and wooded knolls, so characteristic of central and northern Europe, and the northern Mid-West of the USA. In addition to providing good, mineral-rich agricultural land, the deposits are of considerable economic value, providing aggregates for the construction industry and hosting aquifers for underground water supplies.

The most easily identifiable of depositional landforms are 'moraines', since they commonly have long, sharp-crested ridges or ridge complexes, made up of a mixture of glacial sediment ('till') and other deposits pushed up during glacial advances. In a valley setting, the furthest advance of a glacier is marked by a terminal moraine, normally arcuate in form, reflecting the original shape of the glacier snout. Many such moraines, however, are

▲ Glencoe, Grampian Highlands, Scotland
Glencoe is one of the most famous valleys historically in the Scottish Highlands but is also one of the best examples of a glacial trough in the British Isles. Although not truly U-shaped, its parabolic cross-sectional form is more typical of glaciated valleys generally. The valley was eroded out by ice age glaciers, most recently during a cold spell around 11-12,000 years ago. The glacial erratic in the foreground is also a legacy from this time.

▼ **Keswick, English Lake District**
One of the most beautiful areas in the United Kingdom is the Lake District National Park in northwest England. This is a landscape fashioned by ice during the Quaternary ice ages, notably the last one, when ice covered most of the British Isles. These mountains and valleys near the town of Keswick display many glacial landforms, including rock basins containing lakes, although in this image autumn mist hides the lake of Derwentwater.

▲ **Dicksonfjord, Northeast Greenland National Park**
The inner reaches of this region of spectacular fjords once carried outlet glaciers as far as the continental shelf edge, resulting in overdeepened basins and precipitous valley sides. These glacial troughs are the closest of any in the world to having a true U-shaped cross-section.

▼ **An Teallach, Northwest Highlands, Scotland**
Once occupied by a cirque glacier, this ampitheatre of terraced Precambrian sandstones, eroded into the mountain of An Teallach (1062 m /3486 ft) is named Coire Toll an Lochain, and is one of the finest corries or cirques in the Highlands.

▲ **Brüngrat, Berner Oberland, Switzerland**
A knife-edge ridge, referred to as an arête, separates two adjacent glaciated basins. The first basin is a shallow cirque, and is facing the camera, the second is out of sight behind the ridge. The ridge crest is just a remnant of a higher, broader mountain mass.

largely destroyed by meltwater soon after they have been formed. In lowland areas the advances of ice sheets of the past created even larger ridge complexes, that can sometimes be traced for dozens of kilometres. Lateral moraines are deposited by glaciers along the sides of valleys, but hillslope movement and subsequent rockfalls combines to modify them over time. They are most prominent as unstable ridges in areas of recent glacier recession. Many impressive examples were formed during the Little Ice Age (c. 1650-1850). Some actually still retain a core of glacier ice beneath the debris, because of the insulating effect of the debris cover.

There are several other types of moraine from a few to several metres in height, including hummocky moraines, annual push moraines formed by small winter readvances of the glacier snout, and fluted moraines which are long, straight, parallel, smoothly rounded ridges parallel to the ice flow direction.

Some glacial deposits reflect a highly dynamic (fast-flowing) depositional process. If a glacier or ice sheet rides over a plain of till, the material becomes quite wet and easily deformed, and is easily moulded into new shapes. This process leads to the generation of 'drumlins', which are streamlined hillocks of till orientated parallel to the ice-flow direction.

Although glacial transport is best demonstrated by the above landforms, solitary boulders referred to as 'erratics' are also good indicators of the former extent of ice and the direction of flow. Erratics are large blocks of rock that fell on a glacier surface, or were ripped up from the glacier bed. Some have ended up hundreds of kilometres from their place of origin, and were instrumental in helping 19th century geologists develop the theory of ice ages.

Meltwater from a glacier also creates landforms. Wide fluctuations in discharge between summer and winter create braided stream channel systems. The streams modify, sort and redistribute glacial debris, creating 'outwash plains'. In lowland areas these plains may be many tens of kilometres wide, the largest modern examples being found in Iceland, where they are associated with subglacial volcanic eruptions and are known as 'sandar'. Even larger outwash plains are found over central Europe and the American Mid-West, that were produced during a succession of ice ages over the last two million years.

Glacial outwash deposits often bury remnants of stagnant glacier ice. As the ice slowly melts, water-filled hollows, known as 'kettle holes' develop. Beneath the glacier itself channels of water may become choked with debris. As the ice melts, long, narrow ridges of sand and gravel may be left standing, features which are known as 'eskers'. Stream deposits also accumulate along the flanks of a glacier, which are left as isolated benches called 'kame terraces', perched above the valley floor after the ice has melted.

GLACIAL LANDFORMS AND EARTH HISTORY

Erosional and depositional landforms have been intensively studied by geologists in modern glacial environments. By studying the assemblage of glacial landforms, or the 'landsystem' in a particular region, geologists have been able to reconstruct the style and nature of glaciers and ice sheets during former ice ages. From this we know that as much as a third of Earth's land surface was covered by ice during the last and earlier ice ages of the Quaternary Period (the last two-and-a-half million years). Even looking into deep geological time we find evidence that glaciation affected Planet Earth on several occasions, as far back as 2600 million years ago. However, compared with nonglacial periods, periods of glaciation are relatively rare, but they sometimes affected vast areas. Indeed, by recognising distinctive glacial phenomena in the rock record, geologists have inferred that in at least one of these periods of glaciation, around 600-750 million years ago, all the continents may have been covered in ice, an exciting theory that is referred to as 'Snowball Earth'.

◯ **Suilven from Stac Pollaidh, Northwest Highlands, Scotland**
A landscape illustrative of 'deep time'. The lower ground comprises small hills and lakes scoured by glaciers of the last ice age of 20-30,000 years ago – a landscape of aerial scouring. The low-lying hummocky bedrock is Lewisian Gneiss, dating back nearly 3 billion years, while the upstanding mountains are of Torridonian Sandstone, around 800 million years old. The photograph is taken from the popular hill Stac Pollaidh (612 m / 2,008 ft), looking towards Suilven (mid-left; 731 m / 2,398 ft), Canisp (centre; 847 m / 2779 ft), and Cul Mor (right; 849 m / 2785 ft) in the Northwest Highlands of Scotland.

◀▶ **Unterer Grindelwaldgletscher, Switzerland and Corrieschalloch Gorge, Scotland**
Subglacial meltwater is often under high pressure and consequently, with the boulders it carries, it can erode deep 'slot gorges' into bedrock. These two gorges were formed during the last ice age. On the left is the gorge formed below Unterer Grindelwaldgletscher in the Berner Oberland of Switzerland. On the right is Corrieschalloch Gorge in the Northwest Highlands of Scotland.

▼ **Rhonegletscher, Furka Pass, Switzerland**
The effect of glacial abrasion and polishing, as well as bedrock-plucking is evident in these 'roches moutonnées' near the terminus of Rhonegletscher in Switzerland. The granite bedrock shows evidence of ice movement from right to left. Until the end of the 20th century, this area was covered by the glacier, but now an ice-marginal lake has formed behind this rock step.

▶ **Coire Ardair, Grampian Highlands, Scotland**
Hummocky moraine in the glacial valley of Coire Ardair, with the headwall of the corrie (cirque) and the mountain plateau of Creag Meagaidh in the background. These moraines were formed around 11,000 years ago by a small valley glacier, but the main valley was eroded by ice during major ice ages before 18,000 years ago.

○ Canton Zürich / Canton Zug border, Switzerland
A landscape of drumlins and kettle holes. In this region, glacial sediments were laid down during the Last Glacial Maximum around 24,000 years ago. Drumlins are made of till (glacial sediment) that has been streamlined by fast-flowing ice into elongate, smooth hills. Kettle holes with lakes form when ice, buried in sediment, melts out.

▲ St Niklaus, Canton Solothurn, northwest Switzerland
Granite erratics originating in the northern Mont Blanc massif were deposited near the city of Solothurn soon after the Last Glacial Maximum, i.e. soon after 24,000 years BP. These boulders are perched on pedestals of limestone bedrock, which they have shielded from the rain, which otherwise dissolves the rock, during the millennia following their emplacement by the glacier.

▲ Rees River, Southern Alps, New Zealand
Glaciers and snowfields produce abundant meltwater, and the rivers that emanate from them are typically braided as a consequence of constant channel-switching as river discharge varies both daily and seasonally.. The Rees River is one of the many braided rivers that flow eastwards from the snow- and glacier-bearing Southern Alps.

The Legacy of Glaciers & Ice Sheets

◀▼ **Garvellach Islands, Inner Hebrides, Scotland**
The legacy of glaciation is found not just in the recent geological record, but also in the ancient record - in rocks we call 'tillites'. A several-hundred-metre-thick sequence of glacial sediment of late Precambrian age (c. 720 million years ago) is superbly exposed on the uninhabited Garvellach Islands. Almost all the titled strata in this view are of glacial origin, while the close-up view shows the rock's texture. Rocks of this age tell a story of global glaciation, commonly referred to as 'Snowball Earth'.

The Legacy of Glaciers & Ice Sheets   201

CHAPTER 12

LIFE IN A FROZEN WORLD

▶ **St Andrew's Bay, South Georgia, 2019**
A pair of king penguins within a rookery of tens of thousands of these flightless birds, near the coast of this heavily glacier-covered, mountainous island in the South Atlantic Ocean.

For ten thousand years our planet has, until recently, enjoyed a stable climate, allowing plants, animals, and humans to readily adapt to the world around them. The cold regions of the world are home to uniquely adapted plants and animals. However, now many species are under increasing stress from a climate that is warming at an unprecedented rate, reflecting the impact that humans have had since the Industrial Revolution.

Cold regions embrace the land surface covered by glaciers and ice sheets, the surrounding regions of permanently frozen ground ('permafrost'), snow-covered areas, and oceans covered by multi-year and annual sea ice. Species range from microscopic algae and bacteria to the top predator, the polar bear, at Earth's surface, and the great whales in the oceans. All are linked to, and are dependent upon, each other.

ADAPTING TO COLD

Animals and plants have had to adapt to those extensive regions of Planet Earth that are mantled in snow and ice. In perspective, and taking account of both hemispheres in their different seasons, thirty percent of Earth's land surface is covered by snow and ice in an average year, while seven percent of the ocean is covered by sea ice.

On land, all living things have had to learn to cope with deep snow, how to find food and travel in snow, how to survive the fierce winter storms, and then cope with the consequences of rapid melting in spring. Adaptations in mammals include large feet, which help them walk over snow (e.g. hares, ptarmigans) or over sea ice (polar bears), and thick winter fur coats (Arctic foxes) or fat (penguins). Migration is another way that animals cope with changing conditions. For example, some herds of caribou on the mainland of northern Canada migrate northwards from boreal forest to areas of tundra to breed in summer, a round trip of up to 5000 kilometres (3000 miles). Most bird species migrate too, with vast numbers arriving in the Arctic to breed in summer, and then returning to more temperate latitudes in winter. The most extreme example of migration is that of the Arctic tern which nests in the Arctic in summer and then makes a staggering 39,000 km (24,000 mile) round trip each year to spend summer in the Antarctic.

Plants, too, have adapted to sub-freezing conditions in winter. In areas where the ground itself is frozen, trees send their roots down to the unfrozen ground beneath. Standing water allows bogs to develop, providing rich soils that enhance tree growth. These areas are referred to as 'taiga', meaning swampy, moist forest. Permafrost is prevalent in Arctic and sub-Arctic regions. With the ground commonly frozen to depths of hundreds of metres, only the surface upper metre or two, the 'active layer', thaws out in summer. The water-retention capability of frozen ground creates wetlands and inhibits free drainage. The characteristic vegetation of these areas, with an abundance of attractive flowering plants, is known as 'tundra', and trees are absent because the active layer is too shallow to support their roots. In the high Arctic, dwarf tree species, such as Arctic willow and dwarf birch, grow very slowly to only shrub-size and hug the ground.

WILDLIFE IN ANTARCTICA

Glaciers and ice sheets are not totally lifeless, despite the harshness of the environment. The Antarctic Ice Sheet is by far the biggest mass of ice on the planet (as described in Chapter 6). Its highest, innermost parts are the parts of Earth's surface least conducive to life, owing to frigid temperatures, lack of water, and fierce winds. In contrast, at low elevations and towards the edges of the ice sheet, where meltwater can accumulate, algae can survive in ponds, while moss grows in hollows on some nunataks, the islands of rock surrounded by ice. Some bird species, such as the snow petrel, may even fly inland to nest, far from the coast.

The ice-free coastal areas of Antarctica, where adjacent offshore areas provide a rich source of food, are home to large populations of nesting birds. We associate several species of penguin with Antarctica, and it is these that most tourists find most endearing. These flightless birds have adapted to swimming and diving for food, rather than for flying. They are also adapted to coping with sub-freezing temperatures on land or on sea ice in having a thick layer of blubber. All species are able to walk, using their flippers to propel themselves. The iconic emperor penguin

▼ Cierva Cove, Antarctic Peninsula, 2019

Humpback whales accompanied by gulls circling around an iceberg. The whale on the left is 'fluking', showing its tail fin as it begins a plunge to the depths, while the one on the right shows the hump. Sighting of humpback whales is common in the polar regions, now that the hunting of them has largely been abolished.

⊙ Gold Harbour, South Georgia, 2019

One of the richest places for wildlife on Planet Earth is the isolated island of South Georgia in the marine zone between the Southern Ocean and the Atlantic Ocean. Despite being a centre for the exploitation of marine mammals in the past, the concentration of marine mammals and birds is remarkable. In the bay of Gold Harbour, situated below a hanging glacier, we see the dense intermingling of king penguins, elephant seals and fur seals.

Life In A Frozen World

▲ Gerlache Strait, Antarctic Peninsula
A pair of gentoo penguins on a small iceberg. This species is pushing southwards as climate warms, displacing resident Adélie penguins in the process.

breeds on sea ice just offshore from rocky or glacier-covered coasts, with the male seeing the laid egg through the severe weather of winter to hatching until the female takes over in spring. Adélie penguins, another solely Antarctic species, prefer rocky knolls on which to nest, but they spend the winter at sea. More northern Antarctic and sub-Antarctic species, such as the gentoo and chinstrap, commonly climb well-above the ocean to nest. There are many other birds species too, such as petrels, skuas and sheathbills, which often supplement their marine diet with the eggs, chicks, and detritus from penguin rookeries.

The coastal wildlife of Antarctica, including the sub-antarctic islands, is ultimately dependent on the living resources of the surrounding cold Southern Ocean. The keystone species is krill, a small shrimp-like animal, which lives beneath and beyond the sea ice, and this sustains the populations of whales, seals and penguins. Some specific species, the leopard seal and orca (killer whale) in turn feed on penguins, and are often seen cruising amongst ice floes.

TERRESTRIAL WILDLIFE IN THE ARCTIC

Whereas Antarctica is a continent surrounded by ocean, the Arctic is a deep ocean surrounded by land masses with wide continental shelves and numerous islands. Sea ice covers much of the Arctic Ocean and the northern limits of the Atlantic and Pacific oceans, fluctuating dramatically over the seasons. Glacier ice covers many land areas, including most of Greenland and extensive areas in the Canadian Arctic Archipelago, Svalbard and some of the Russian islands. Beyond the glacier-covered regions, a long winter season means that snow can last for well over half the year, while the ground remains frozen, as permafrost, to depths of hundreds of metres (up to a thousand feet).

▲ **Northern Antarctic Peninsula**
Everyone who visits Antarctica is keen to see the several species of penguin that breed there. The most common species in the northern Antarctic Peninsula are chinstrap and gentoo penguins.

The glaciers and snow slopes of the Arctic have become a breeding ground for algae and bacteria, which stain the surface pink and green. These micro-organisms also collect around sediment particles in 'cryoconite holes' on the ice surface, features which are formed by enhanced melting around dark objects. The role these micro-organisms play in increasing the melting of glaciers has only begun to be recognised in the last two decades.

Among the Arctic icefields we find that sea birds, including Arctic skuas, ivory gulls and fulmars, fly inland to nest on nunataks, as well as to coastal cliffs. Where birds are nesting, there is also a good chance of finding predatory Arctic foxes.

In areas adjacent to the glaciers, the ground is permanently frozen, yet the thawing out of the top one or two metres in summer provides ponds and supports a delicate tundra vegetation. Mosses and lichens soon colonise ground that has been vacated by glaciers, and pioneer flowering plants also begin to brighten the bare ground. On ground exposed for centuries or more, the tundra may only be patchy, but in summer the wide variety of Arctic flowers are a delight to the eye, examples being saxifrages, Arctic poppies, moss campion, mountain avens, as well as mosses and grasses. Ground-hugging tree species include Arctic willow and dwarf birch which, with berry-bearing shrubs, put on a glorious display of colour in the autumn. The tundra, ponds and adjacent crags support vast numbers of breeding birds in summer: geese, ducks, gulls, terns, skuas, guillemots (known as murres in North America), puffins, snow buntings, and ptarmigans, amongst others.

The land areas of the Arctic support several mammal species, including reindeer in Svalbard and northern Eurasia and the closely related caribou in the Canadian Arctic. Greenland, the Canadian Arctic, and Alaska, provide a home for Arctic hares, musk oxen and rodents, as well as their predators, foxes and wolves. Because of hunting, now fortunately much reduced, few of these animals are dangerous to people. However, from personal experience, a solitary musk ox, may be sufficiently intolerant to charge a human if it is disturbed inadvertently.

MARINE WILDLIFE IN THE ARCTIC SEAS

Most offshore areas in the High-Arctic freeze over in winter, the main exception being where the warm waters of the North Atlantic extend up to western Svalbard. In the Arctic Ocean, centred on the North Pole, the sea ice does not melt completely in summer, and multi-year ice forms.

Life In A Frozen World

However, the area of year-round ice is shrinking dramatically as a consequence of global heating, and many sea-ice scientists are predicting an ice-free Arctic Ocean in summer within just a few decades. Already sea ice thickness and extent have declined to levels not seen in historical time.

Arctic waters are rich in marine wildlife, and here we find a complex food web, within which micro-organisms and higher species all interact, and depend upon each other. Thus, crustaceans and fish support species of seal, walrus and whale, while seals sustain the polar bear. Indeed, biologists consider the polar bear to be a marine species, since this species uses sea ice as a hunting platform for seals. Often portrayed as the 'King of the Arctic', the polar bear is an apex predator, sitting at the top of the food web. Sadly, the main threat to this iconic species, that all visitors to the Arctic want to see, comes from humans themselves. Until the 1970s, the threat was from sports hunting. Then, hunting was banned by international treaty, but more recently pollutants and sea-ice melting have been having a negative effect on their long-term breeding success. Travellers over land in the Arctic always have to be prepared for a polar bear encounter by carrying flares and a rifle, but by careful surveillance of the landscape and avoidance, such an incident can normally be avoided. The best sightings are thus usually at sea from the protection of a ship or a Zodiac-style boat.

WILDLIFE OF THE HIGH MOUNTAINS

Climatic conditions on the highest mountains on Earth are similar to those in the polar regions, even those lying in temperate latitudes. As in the polar regions, the most primitive organisms are algae and viruses, which grow on snow and ice. On glaciers, beneath stones, even insects can thrive. However, because of extreme altitudinal variations, the coverage of

▲ **Gerlache Strait, Antarctic Peninsula**
A close encounter with the most ferocious of marine mammals, the orca or killer whale. Photograph taken from a Zodiac inflatable boat.

▲ Hornsund, southwest Spitsbergen, Svalbard
Humpback whales follow our ship at the mouth of the southernmost fjord in Spitsbergen.

snow and glacier ice is extremely variable, as are the associated ecosystems. Many of the world's temperate mountainous regions receive copious amounts of snow and support glaciers. As the glaciers retreat, plants rapidly colonize the newly exposed ground, as the sediment that becomes exposed is commonly rich in minerals. For example, in the European Alps pioneer plants first arrive within a few years, having been transported as seeds by wind or birds. These early plants include saxifrages and grasses, and may often be found within a couple of years of the ice receding. Decaying organic material from these pioneer plants then prepares the ground for more demanding species, such as the larger flowering plants and shrubs, including willow, alder, birch and *Alpenrosen*. The colonization sequence continues with the growth of larch and is complete when pines trees begin to take over.

At high altitude, we find ice-free ground that is bare of vegetation and quite likely underlain by permafrost. Moving downhill, we pass into alpine meadows, rich in wild flowers, and then into heavily forested zones. Snow deeply buries all of this terrain in winter.

The vegetation provides food for herbivores. In the Alps stone ibex and chamois are common, scrambling with ease over precipitous rocky hillsides, while marmots emerging from holes in the ground make a whistling sound to warn their mates at the first signs of danger. Less common are wolves and foxes, but making a come-back in some areas after centuries of persecution. Aloft, soar birds of prey, such as eagles, waiting for opportunity to swoop on unsuspecting rodents below. In the mountains of western North America, Dahl sheep and mountain goats inhabit terrain similar to that of the ibex and chamois, whilst in less rugged terrain moose, elk, bears, and other animals frequently forage for food.

Life In A Frozen World

WILDLIFE OF CONTINENTAL INTERIORS

Beyond the mountains, in the continental interiors of North America and Eurasia, winter snow may lie long and deep. Much of this terrain has been heavily influenced by humans, and the natural distribution of animal and plant species has been altered. Conflicts between mammals and humans are common, but in some protected regions, native lowland animals still roam freely, familiar examples being deer, bison, hares, rabbits, and rodents, as well as their predators red foxes, wolves, coyotes, and bears.

HUMAN IMPACT ON COLD-REGION ECOSYSTEMS

In summary, snow-covered regions demonstrate that plants and animals can thrive where humans cannot. Visitors who venture into such regions may be privileged to see a remarkable variety of life, well-adapted to extreme environments. However, over the centuries, humans have devastated many parts of the world through industrialisation and uncontrolled hunting. Permafrost areas have been degraded, and populations of seals, whales, polar bears, and foxes, to name just a few species, were once on the verge of being made extinct. Thankfully, protection is now enforced in many areas.

Now, however, there are more insidious threats – global heating and pollution. The extent of land ice, snow and sea ice is shrinking rapidly, and at an accelerating rate. Temperatures in the Arctic, Antarctic Peninsula and some mountain regions are rising at several times the global average. Animal species are being displaced from their usual habitats, from polar bears and sea birds in the north, to emperor and Adélie penguins in the south. Areas of permafrost are decaying, fundamentally changing ecosystems, and releasing methane into the atmosphere. Methane is a much more potent greenhouse gas than carbon dioxide, but quantifying its release into the atmosphere is much more difficult to predict. Wildfires are occurring with increased frequency as a consequence of rising temperatures and increasing periods of drought. The urgency to tackle the climate crisis is ever greater, for without concerted action, the consequences for the biosphere will be disastrous.

▲ **Cuverville Island, Antarctic Peninsula, 2017**
Pink algae covers the surface of a snowbank on a relatively warm rainy day in late January. Temperatures in the Antarctic Peninsula have been rising rapidly in recent decades. Periods of rain encourage algal growth, but conversely have an adverse effect on nesting birds, especially penguin chicks.

▶ **Skalafellsjökull, southern Iceland**
Lichens and mosses on a glacier forefield.
(top) Among the first plants to colonise the deposits in front of a receding glacier are colourful lichens and bright green mosses.

(bottom) Close-up of the orange, black and green lichens on a glacially transported boulder.

213

A selection of flowers which grace the Arctic tundra in summer-time, with examples from the Canadian High-Arctic and Svalbard.

▲ *Purple saxifrage, Longyearbyen, Svalbard*   ▼ *Moss Campion, Kongsfjorden, Svalbard*   ▲ *Arctic mouse-eared chickweed, Axel Heiberg Island, Nunavut*

214   Our Frozen Planet

▼ *Arctic poppy, Axel Heiberg Island, Nunavut*   ▲ *Alpine arnica, Axel Heiberg Island, Nunavut*   ▼ *Moss Campion, Kongsfjorden, Svalbard*

▶ **Ingmikêrtikajik, Scoresby Sund, East Greenland**
The High-Arctic is a zone characterised by tundra that lacks typical trees. However, ground-hugging shrubs such as arctic willow (yellow) and alpine bearberry (red) form impressively colourful displays in the autumn (September), as on this small island in Scoresby Sund.

▲ **Bylot Island, Nunavut, Canada**
Arctic willow grows well on the open tundra at low elevations on Bylot Island, especially where it can find shelter in the lee of boulders.

▲ **Arctic Ocean**
Female polar bears give birth to their cubs in snow caves in winter. After bringing them out into the open in spring they hunt with their cubs on the sea ice for over a period of at least two years completely independent of the father. This mother (centre) has a pair of 2-year-old cubs crossing sea ice in the Arctic Ocean, several hundred kilometres from the nearest land at Franz Josef Land.

◂ **Nordaustlandet, Svalbard**
Walrus resting on an ice floe seaward of the ice cap of Austfonna on the east side of Nordaustlandet (Northeast Land). The ice cap delivers icebergs into the ocean, several of which are seen in the background.

▸ **Ella Ø, Northeast Greenland National Park**
An Arctic hare standing next to a glacial erratic on Ella Ø, a beautiful island in Kong Oscar Fjord. These animals remain white throughout the summer and are quite conspicuous in the landscape at that time. Arctic hares are resident only in the North American High-Arctic.

Life In A Frozen World

◀ **Ella Ø, Northeast Greenland National Park**
A solitary musk ox roaming freely over this small but mountainous island. Normally in herds of several animals, musk oxen graze on the tundra vegetation. They are capable of surviving a wide range of temperature conditions due to their shaggy coats which become matted and straggly in summer. Their wool is prized for its exceptionally high insulation quality.

◣ **Ny-Ålesund, northwest Spitsbergen, Svalbard**
The Arctic fox is well-adapted to Arctic climates, having short greyish brown and fawn fur in summer, and a thick white coat in winter. The white fur has given rise to extensive hunting, although the species is at least partially protected nowadays. Foxes are often found feeding on eggs and young casualties below bird cliffs in the breeding season, but can also be found foraging or sheltering around human settlements.

▶ **Color Lake, Axel Heiberg Island, Nunavut, Canada**
An Arctic wolf ambles undeterred by visiting glaciologists along the shore of this lake at the base camp of the McGill Arctic Research Station.

Life In A Frozen World

◀ Morteratsch valley, Engadin, Switzerland, 2002 and 2018
As these repeat photos show, pioneering plants quickly colonize terrain after Vadret da Morteratsch, the glacier in the background, has receded from this location. Of the perennial plants, willow (on the left) is the first to appear, followed by larch (right) and then, somewhat later, pines.

▲ Alpine flower portfolio, Switzerland
Some of the species of alpine flowers that grace the snowy high mountains and meadows. Top row: alpine toadflax; cobweb house-leek; spring gentian. Middle row: moss campion; alpine snowbell; yellow mountain saxifrage; Bottom row: Scheuchzer's bellflower or Campanula; alpine mouse-ear chickweed; hairy alpenrose

Life In A Frozen World 223

◀ **Obers Ischmeer near Grindelwald, Berner Oberland, Switzerland**
Stone ibex are particularly well-adapted to life at high altitude, and glaciers. When the summer heat becomes difficult for them to bear, small herds seek out old snow patches on which to rest. Suitable vegetation exists above and beyond Little Ice Age limits as here, high above the glacier named Obers Ischmeer, but the animals have no hesitation crossing glaciers if they need to.

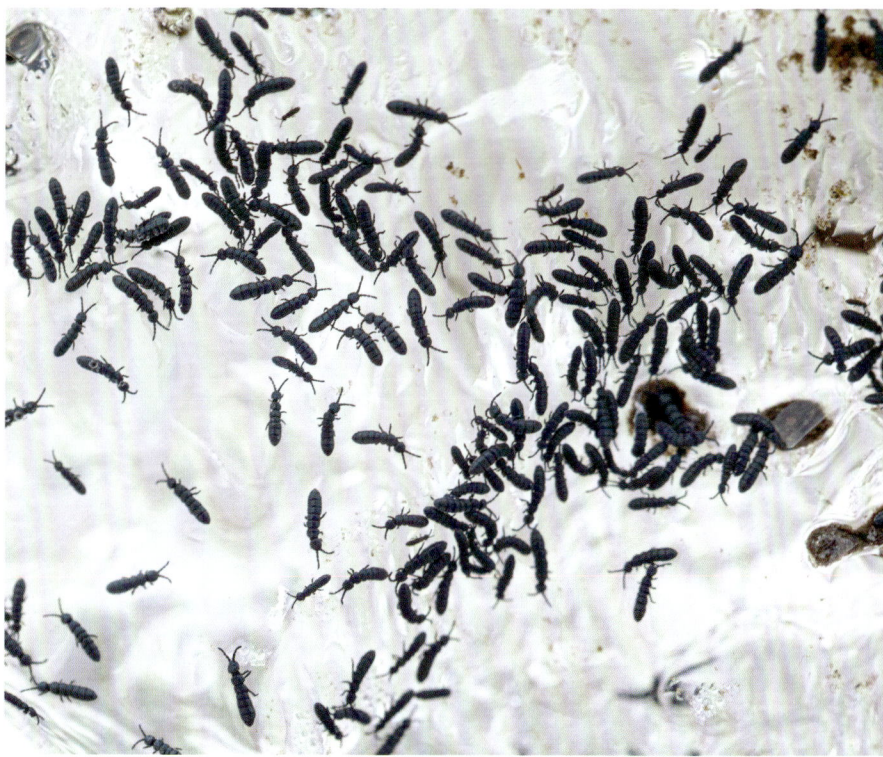

▲ **Vadret da Morteratsch (Morteratsch Glacier), Engadin, Switzerland**
Springtails, colloquially known as 'glacier fleas', thrive in cold conditions. While these primitive insects die at temperatures higher than 10°C (50°F), they are perfectly comfortable on melting snow or glacier ice, often congregating under boulders on the ice.

Life In A Frozen World

▼ Yellowstone National Park, Wyoming, USA
Bison grazing near Old Faithful, the most famous of geysers in the National Park. In order to reach vegetation, bison swing their massive heads back and forth, removing the snow as though with a giant brush.

▶ Alaska Range, Alaska, USA
The 'taiga' is the zone of vegetation between dense coniferous forests and open tundra. Taiga is dominated by marshy ground and slender trees that shed snow more readily than those further south. The wet ground with ponds in summer is a consequence of the underlying permafrost inhibiting drainage. This scene is viewed from a train on the Alaska Railroad.

CHAPTER 13

SNOW, ICE & SOCIETY

◂ **Flyvefjord, Northeast Greenland National Park, 2018**
Some of the majestic icebergs that calve from a massive outlet glacier of the Greenland Ice Sheet into Nordvestfjord have become stranded in Flyvefjord, a branch of the main fjord. Specialist polar cruise ship operators give tourists the chance to see some of these wonders of nature close-up, and to learn to appreciate how climate change is so evident in the polar regions.

Human societies have used and adapted to snow and ice for thousands of years. The indigenous people of Alaska, northern Canada, Greenland, northern Scandinavia and Siberia are true masters of survival in cold climates, where temperatures regularly drop below -50°C (-58°F), and where a snow-covered winter landscape is normal. For these people, it is the summer months that bring the greatest challenges. The inhabitants of northern Siberia, for example, find that the brief warm summer is the worst time of the year, since travel over water-logged, swampy terrain is much more strenuous than sledging over snow. Frost and snowfall in the early autumn bring respite to humans and their reindeer herds from the misery wrought by the swarms of mosquitoes that breed in swamps in summer. By contrast, in mid-latitudes, humanity finds snow and ice a hindrance or even a hazard, but also offering opportunities for enjoyment and enhancing life-styles. In this chapter we will explore some of the benefits of snow and ice, as well as their adverse effects on humanity.

ICE AND SNOW HARVESTING

Before the invention of artificial refrigeration, snow and ice played an important role in the preservation of foodstuffs. Ice was harvested from frozen lakes and even rivers. In the cold winters of the Little Ice Age, between 1600 and 1815, the River Thames regularly froze over in London. Several major 'frost fairs' and dozens of small ones were held on the ice, with temporary shops, pubs and skating rinks serving the needs of Londoners. The ice itself was used for refrigeration – broken off in large slabs and then hauled ashore.

Elsewhere, river and lake ice was transported over much greater distances, commonly by rail, barge or ship. Ice houses were built to store ice for later distribution and domestic consumption. Ice-harvesting from rivers grew into big business in the United States in the 19th century, revolutionising the meat, fish, vegetable and fruit industries, and employing tens of thousands of people. Trading of this natural ice grew between several countries, for example with Norway exporting ice to Britain. By 1900, the ice trade was declining, and was essentially extinct soon after the First World War.

In the absence of river or lake ice, deposits of wind-packed snow were exploited instead. Harvesting of snow was practiced in China as early as 1000 BC, while the ancient Greeks and Romans insulated pits with grass and branches of trees, then filled them with packed snow in order to cool beverages during the hot season. Even glacier ice was used in great quantities for refrigeration. Ice cut from Glacier de Trient in the Mont Blanc area was first moved a few kilometres by horse-drawn, narrow-gauge wagons to Col de la Forclaz, a mountain pass. It was then carried on the back of mules down steep paths to the town of Martigny, from where it was exported by rail to Paris.

Rather than carrying the ice to the place where refrigeration was needed, sometimes the process was reversed, by moving foodstuffs to natural locations that remained cold throughout the summer. Milk produced on Alpine pastures often had to be stored for some time before it could be transported down to the valley, or before cheese could be made. Milk storage in ice caves was common in the Jura Mountains along the border between France and Switzerland. Nowadays, of course, we no longer depend on natural snow and ice for refrigeration.

MELTWATER

Snow-packs and glaciers provide a ready source of water in mountain regions, especially during dry seasons. The Himalayan region has the largest concentration of glaciers outside the polar regions, covering 33,000 sq km (12,741 square miles). Given the region's proximity to the homes of billions of people, it is sometimes aptly referred to as the "Water Tower of Asia". It delivers 86 million cubic metres (19,000,000,000 UK gallons / 22,500,000 US gallons) of water to some of the largest and most heavily used rivers in the world. The seven largest of these rivers are the Ganga, Indus, Brahmaputra, Salween, Mekong, Yangtze and Huang Ho, and the melting ice ensures a year-round water supply to about one billion people. However, recent years have seen an acceleration of glacier recession as a consequence of global heating, and the prognosis

▲ **Monlesi ice caves, Jura Mountains, Switzerland**
Ice caves such as the 'Glacière de Monlési' in the Jura Mountains of Switzerland remain cold throughout the summer and in historic times were therefore commonly used for storing perishable agricultural produce such as milk. The ice forms because cold, dense winter air flows from a high entrance into a low cavity and remains trapped there when temperatures rise outside.

◀ **Lej da Silvaplauna, Engadin, Switzerland**
Sports on snow and ice are among the benefits of life at high altitude or in cold climates. Curling combines the skilful use of the unique properties of ice and complex strategic thinking. These players are practising next to the lake of Lej da Silvaplauna, near the famous resort of San Moritz.

▶ **Kaunertal, Tyrol, Austria**
Some glaciers in the European Alps have been used for summer downhill skiing. Pylons for the associated ski-lifts can be anchored directly in the ice. This resort in the Tyrolean Alps is suffering from glacier recession. In the foreground blankets protect reserves of piled-up snow which will be used in the following summer to reduce the loss of glacier ice.

◀◀ **Luosto, Lapland, Finland**
A traditional way of travelling long distances on snow-covered terrain makes use of reindeer, rather than horsepower. This reindeer is so accustomed to taking tourists on a circular track that it will do so without an accompanying driver.

◀ **Grindelwald, Berner Oberland, Switzerland**
During an annual snow festival in the mountain resort of Grindelwald, groups compete in producing a variety of snow sculptures. Depending on the weather, these works of art may last only a few days, which nevertheless is long enough to involve the public in the judging.

for these glaciers on the time-scale of the next half century is poor. Serious consequences thus lie ahead for the freshwater ecosystems of the river basins, with long term impacts on biodversity, people and livelihoods, as well as on regional food security.

Many countries with high mountains are able to generate much of their own energy from hydro-electric power, to which snow-melt and glacier-melt are major contributors. For example, Switzerland has, arguably, captured more water for hydro-power than any other country. At least 680 hydro-schemes, involving numerous tunnels and dams, provided 61.5% of the electricity used in that country in 2021. Switzerland also has a network of irrigation channels, often perched on steep mountainsides, that are used for agriculture. Such channels are a feature of many populated mountain regions; the Cordillera Blanca in the Peruvian Andes is another prime example of this type of usage.

WINTER SPORTS AND HOBBIES

General improvements in living standards have given us the opportunity to utilise water in its frozen state for a whole range of sporting activities, notably skiing. In the upper reaches of some Alpine glaciers, snow remains accessible by cable car throughout the summer. Such destinations not only offer hikers a welcome respite from high summer temperatures in the valleys, but also allow summer skiing. However, glacier recession is increasingly making maintenance of the summer skiing facilities difficult and cost-intensive. Operators of such facilities also have to accumulate large piles of snow in winter, and redistribute it onto pistes in summer as required. Additional snow also has to be produced artificially using snow cannons, but this requires the construction of reservoirs, considerable energy expenditure, and has negative consequences for alpine vegetation.

Downhill skiing and snowboarding are the most important drivers of the tourist economy in the Alps, the North American Cordillera and similar mountain areas. Revenues from winter tourism far exceed those of summer. Climate change and related rising air temperatures are, therefore, a serious threat to the mountain tourist industry, particularly in middle latitudes. Smaller resorts, particularly those at modest altitudes, increasingly find snowmaking too expensive, and the skiing season too short, to remain economically viable. There is increasing focus on large, high-elevation resorts, resulting in longer distances people have to travel. Ironically, this results in increasing greenhouse gas emissions and other pollutants which, paradoxically, help destroy the remaining resources of snow and ice.

Other forms of winter sports, such as cross-country skiing, snow-shoeing, ice-skating, curling, dog-sledging, reindeer-sledging are other enjoyable ways to experience snow and ice. In particular, in the colder northern countries, cross-country skiing is very popular, and is not dependent on high relief. As winters have become milder in Central Europe and the United Kingdom, for example, Scandinavian destinations have grown in popularity.

In a few places, a unique tourism experience can be found in ice hotels, where guests can sleep in rooms entirely made of ice. The largest example of this kind can be found at Jukkasjärvi in northern Sweden, which uses natural ice from the nearby River Tarne. Other ice hotels are to be found in Norway, Finland, Austria and Canada. The northern regions have an added attraction in the form of the Aurora Borealis or Northern Lights, which are enhanced by the snowy whiteness of the landscape.

In regions of the world where longevity of snow and ice is anticipated, festivals have sprung up in towns and villages. The most famous is the Harbin International Ice and Snow Scupture Festival in China, but even small communities go all-out to attract visitors to view their ephemeral works of art.

SHIP-BASED TOURISM

Another form of tourism, which seeks out snow- and ice-covered terrain for its scenic and wildlife value, is on ship-based 'expeditions' to the Arctic, Antarctic and the western fjords of North America and Chile. Typically, specialised ice-strengthened expedition ships, carry passengers to places that were formerly inaccessible to most people. Using a fleet of rugged rubber boats, tourists are taken ashore at designated locations for short walks with knowledgeable guides. Wealthy tourists, in the recent past, were even able to take a voyage to the North Pole on a nuclear-powered Russian icebreaker. Recession of sea ice has made voyages to Svalbard or northern parts of Greenland much easier during the past decade than in earlier years, and even a circumnavigation of Svalbard nowadays is possible in most summers, something unheard of a few decades ago. Guests on expedition ships experience some of the most outstanding natural scenery on Earth – a magical world of calving glaciers, icebergs, deep fjords, spectacular geology, colourful summer vegetation and wildlife. Top of the wish-list for most visitors, however, are encounters with wildlife, such as polar bears in the Arctic and penguins in the Antarctic.

GLACIER TOURISM

Nearer to centres of population, some glaciers remain accessible in summer, providing the visitor with an impression and insight into the fascinating world of snow and ice. In glacier tongues, such as those of Rhonegletscher in the Swiss Alps and Mer de Glace in the French Alps, tunnels have been cut into the ice, enabling the tourist to experience the inside of a glacier. Outside, guided walking tours and ice-climbing activities take place on New Zealand's best-known glaciers, the Fox and Franz Josef, using helicopters to uplift visitors to the best and safest areas for such activities. Scenic flights of these and other glaciers and peaks in the Aoraki/Mount Cook region are also popular in helicopters and small fixed-wing aircraft.

On the other hand, other glacier tongues previously used intensively for tourism, have quite simply melted away. For example, Grindelwald in the Bernese Oberland in Switzerland, formerly known as the 'glacier village', used to have a view of, and easy access to, the Unterer and Oberer Grindelwaldgletscher, but this is no longer true. Many hours of walking are now necessary in order to view their recessed tongues, while direct access is now considered to be dangerous. Consequently, the Alps, most famous for their 'eternally' snow-covered summits, are in the process of losing their decorative white adornment in summer for good. These changes are already having an impact on mountaineering, with some peaks, including the Matterhorn becoming too dangerous from rockfall during the summer. Indeed, traditional Alpine tourism in general is under threat, and communities are having to provide new facilities to maintain their economies.

WEATHER-RELATED HAZARDS

So far, we have emphasized the many benefits of snow and ice for humanity. These, however, come at a price. Large accumulations of snow may result in avalanches (Chapter 2), which are one of the most common natural dangers in any mountainous area. Avalanche danger for centuries has required careful selection of locations where settlements could be built safely. Forests played a vital part in effectively protecting Alpine villages from the destructive forces of avalanches. A growing population, along with roads and railways, require additional protective measures. Extensive arrays of avalanche fences and long avalanche galleries require considerable investment during construction and maintenance. Despite all this, some mountain villages that are usually accessible by car or rail may be cut off during periods of severe snowfall for several days or even weeks.

Snow and freezing rain can also wreak havoc on transport and infrastructure in less mountainous regions. Alpine and Scandinavian countries benefit from an established technical infrastructure for snow removal, such that roads and railways are snow-ploughed, roads are treated with sand, while car owners routinely change their tyres between summer and winter seasons. In contrast, the authorities in regions with only intermittent snowfalls see similar investments as uneconomic, so when a large snowfall or freezing event occurs, the disruption is considerable. Salting or gritting roads provides a partial solution, but even a few centimetres (or an inch or two) of snow or freezing rain may lead to a complete standstill of road and rail traffic.

GLACIER HAZARDS

Glaciers have played an important role in natural disasters and, in a few cases, have caused the loss of thousands of lives. Active glacier-covered volcanoes are notoriously dangerous. In Iceland, subglacial volcanic eruptions are fairly common and melt enormous amounts of glacier ice. The meltwater eventually drains along the glacier bed and causes massive flooding in the adjacent lowlands, events referred to as 'jökullhlaups'. As such events can be forecast well in advance, loss of life nowadays can usually be avoided by timely closure of roads across the flood plains. An eruption of the volcano Eyjafjallajökull in Iceland in April 2010 was one event that caused extreme travel disruption. The interaction of hot lava with glacier meltwater produced a huge plume of ash that led to the closure of air space over much of Europe. Millions of people in Europe and beyond were stranded as flights were cancelled, due to the concern that ash could damage aircraft engines.

Much more tragic was the outcome of a relatively minor eruption of volcano Nevado del Ruiz in Colombia on 13th November 1985. Despite warnings from geologists that volcanic mudflows called 'lahars', resulting from mixing of meltwater and debris, were a likely consequence of the anticipated eruption, no action was taken to evacuate people living around the volcano.

▲ **Søbotn, Troms, northern Norway**
Increasingly, tourists from temperate latitudes wish to experience a classical winter wonderland which, due to climate change, becomes less possible near their home. Consequently, many head to subpolar or even polar latitudes where cold winters are still common. An added benefit to such a journey is the aurora borealis here seen above a holiday home in the Troms region.

Thus, when the eruption took place, unnoticed because of bad weather, lahars began sweeping down the flanks of the volcano. Within minutes of the eruption, hundreds died in the villages closest to the volcano. Even worse, one particularly large lahar travelled much further. Without receiving an effective warning, the city of Armero was overwhelmed by the lahar in the middle of the night, and three-quarters of the 29,000 inhabitants perished.

Mudflows have also repeatedly cost lives in the Peruvian Andes, but these are not usually related to volcanism. Glaciers in Peru have been receding massively since the mid-20th century, leaving behind ice-cored terminal moraines. Continued recession allows proglacial lakes to form, dammed by these unstable moraines. Failure of the moraine dam results in a 'glacial lake outburst flood', and many have been recorded, some with considerable loss of life. As an example, on 13th December 1947, the moraine-dam retaining the small Laguna Cohup burst, resulting in a mudflow that killed 5000 people in the city of Huaraz. Since that tragedy, Peru has made considerable effort to identify newly formed glacial lakes and, where possible, to drain such lakes by carefully lowering the moraine dam or by tunnelling through bedrock to extract excess water. Similar procedures are used in the Himalaya, where the outburst flood risk is also high, but lifting the necessary equipment to high altitude in countries like Nepal (upwards of 4000 metres) is a serious challenge.

Returning to Peru, the worst catastrophe involving a glacier in that country's history took place during the great earthquake of 26th May 1970, when roughly 50 million cubic metres of ice and rock broke off Nevado Huascaran Norte, a 6652 metre (21,824 ft) high peak not far from Huaraz. A mixture of ice, meltwater and debris took just a few minutes to travel 16 kilometres (10 miles) downslope towards the town of Yungay, where 18,000 people died.

SCIENTIFIC RESEARCH ON SNOW AND ICE

The scientific study of ice and snow is referred to as 'Glaciology'. The first written accounts of glaciers date from the 12th Century in Iceland, and the first links to climate where made as early as the following century. Icelandic scientists had a good understanding of their glaciers by the time geologists and physicists began studying glaciers in the Alps in the late 18th century, often as a by-product of their mountaineering exploits. The science of glaciology and the development of 'Ice Age Theory' evolved through the field observations and sharing of knowledge by geologists and physicists from several European countries. Understanding of snow and ice grew through the 19th century in parallel with that of geology and biological evolution, and by 1936 a group of glaciologists decided to found the British Glaciological Society, an organisation that soon adopted an international outlook. This was somewhat ironic for a nation lacking modern glaciers, but later the organisation morphed into the International Glaciological Society that we know of today. The growth of the subject has been exponential since World War II, with increasing emphasis on the polar regions. With the recognition of the importance of the cryosphere on the health of our planet since the 1980s, glaciology is now regarded as a mainstream science. Research is undertaken by groups in many universities and government institutes worldwide. Nowadays, there are many organisations that focus on glaciology, such as the World Glacier Monitoring Service, the National Centres for Snow and Ice, the Swiss Federal Institute for Snow and Avalanche Research, and several governmental polar research surveys, to name just a few.

Glaciology is a multidisciplinary science that draws, for example, on geology, geography, physics, chemistry, biology and mathematics. Recent major advances in understanding of snow and ice have come from the use of satellites and numerical modelling. Glaciology is essential for understanding how humans are affecting the climate and the resulting global consequences. However, it is still fieldwork that underpins everything we need to know about the role of snow and ice in the Earth System. Fieldwork may involve well-funded teams setting up large-scale

infrastructure in the middle of the vast ice sheets, or sending ships to study sea ice changes through the Arctic Ocean's polar night. But it has often been through small teams with limited resources, working in some of the remotest and most beautiful areas of the planet, that some of the major advances in understanding the cryosphere have been made.

IMPACT OF SNOW AND ICE ON SOCIETY

There is a long list of negative and sometimes catastrophic impacts from snow and ice on humanity, such as glacial lake outburst floods, mudflows and avalanches, declining water resources, especially for people living in high mountain regions. Globally, reduction in glacier ice cover is raising sea levels, and the decrease in sea ice is affecting weather systems. For many communities, however, the existence of snow and ice in general is beneficial. Snow and ice provide an all-important source of fresh water, particularly in times of drought or in countries that have a dry season. Snow also provides a playground for humans to enjoy physical activities, that every four years culminate in the Winter Olympics. However, the continuing and accelerating decrease of snow-cover and sea ice, and the recession of glaciers, is already having a detrimental effect on many ecosystems and, indeed, on human civilisation itself.

▲ Orne Harbour, northern Antarctic Peninsula, 2019
Visiting the snowy and icy polar regions is an increasingly popular touristic experience, especially from the comfort of an ice-strengthened expedition ship. This vessel, MV Ocean Adventurer operated by Quark Expeditions, is viewed from a Zodiac inflatable boat in a glacier-fringed bay on the west coast of the Peninsula.

▲ Gåsefjord, East Greenland, 2016
Most expedition ships have a fleet of Zodiac or equivalent inflatable boats to give passengers a chance to experience ice at close quarters, as this scene of bergy bits in Gåsefjord, the southernmost branch of Scoresby Sund, shows.

◀ **Franz Josef Glacier, Southern Alps, New Zealand, 2006**
A guided tour across a glacier is a unique experience for most people. The surface of this glacier used to be reached safely from the snout by traversing through regions of melted out crevasses, as this party of Aberystwyth University (Wales) students was doing. Rapid recession of the glacier and stone-fall now precludes safe access, except by helicopter.

▶ **Pontresina, Engadin, Switzerland**
Villages, roads and railway lines in the Alps are prone to damage from snow avalanches. Counter-measures include the installation of arrays of fences which prevent the snow on steep slopes from sliding off, and protects the tourist town of Pontresina.

CHAPTER 14

A FAREWELL TO SNOW & ICE

▶ **Wormald Glacier, Antarctic Peninsula, 2012**
As in most parts of the world, glaciers in the Antarctic Peninsula are thinning and receding. The British scientific station of Rothera is located on Adelaide Island on the western side of the Antarctic Peninsula, and provides magnificent views of the surrounding glaciers and mountains, including the ice cliffs of Wormald Glacier.

We hope this book has emphasised to our readers the beauty and majesty of snow and ice; how this simple compound of hydrogen and oxygen in its solid state continues to shape our planet; and how it supports humanity and wildlife. We hope, too, that we have managed to convey how snow, sea ice, river ice, lake ice, glaciers and ice sheets are all shrinking under the impact of human-induced climate change - and why this is critical for the future of humanity and the biosphere.

THE GREENHOUSE EFFECT AND HUMAN IMPACT ON CLIMATE

Life as we know it is dependent on the greenhouse effect, which is simply the warming process that occurs when some gases in Earth's atmosphere trap solar radiation. Although 99.6% of Earth's atmosphere is made up of oxygen and nitrogen, these gases do not trap solar radiation. Rather, it is the remaining 0.4%, principally carbon dioxide, along with water vapour, methane and nitrous oxide, the so-called greenhouse gases, that traps the solar radiation. It is these gases that makes our planet much warmer than it would be otherwise.

However, as humans are now adding these greenhouse gases into the atmosphere at geologically unprecedented levels, notably by the burning of fossil fuels, by the destruction of forests and peatlands that store carbon, and through agricultural practices. Indeed, the carbon dioxide level in the atmosphere, as measured at Mauna Loa in Hawaii, had reached 426.9 parts per million by May 2024, and was still rising. According to the US Government agency NASA, human activities have raised the carbon dioxide content of the atmosphere by 50% in less than 200 years. Increasing greenhouse gases means that our planet is getting hotter. Globally, temperatures had already reached 1.4°C above pre-industrial levels by 2023. Furthermore, on current (2024) trends, temperatures are set to rise to nearly 3°C above the pre-industrial average by 2100. Within a geological context, the 10,000 year-long 'Holocene Epoch' of stable climate, during which human civilisation has evolved, and the glacial/interglacial cycles of at least the last one million years, may have been shattered by the temperature rises caused by recent human activity. The dramatic increases in greenhouse gases are the fundamental reason for the continuing and accelerating shrinkage of snow and ice.

MEETING TARGETS FOR GREENHOUSE GAS REDUCTION

Geologists look to the past climate and associated environments to inform us about what might happen in the future – reversing the old geological adage "The present is the key to the past" to "The past is the key to the future". Thus we envisage a world, with few glaciers, ice sheets much reduced in size, sea levels several metres higher with coastlines shifted inland, and quite different climatic and vegetation regimes – a world that is heading our way!

Climate scientists, through the UN's Intergovernmental Panel on Climate Change, tell us that we need to keep global temperatures down to 1.5°C above the pre-industrial average, beyond which Earth will experience catastrophic climate changes. Urgent action is needed to meet this target, yet it is predicted that we will pass this level within 5 years (from 2024). We are thus faced with a 'climate emergency', and while it is the developed industrial countries that have largely caused the problem, it is the less-developed countries that will suffer the most. Indeed, flood risk is increasing in all coastal regions from rising sea levels, water resources are declining, climatic belts are shifting, more extreme weather events are occurring, forests are burning and permafrost is melting – all as a consequence of global heating.

A landmark agreement in Paris in 2015, committed nearly all the nations of the world to reducing greenhouse gas emissions, but progress since has been too little and too slow. Consequently, out of concern for the inadequate action to tackle climate change, young people, along with others, have been 'striking for climate' and demonstrating that 'business as usual' is not an option. Similarly, aid and conservation organisations have been pressing governments and local authorities to act, and promise to become carbon-neutral by 2050 or sooner. The first national governments to declare a 'climate emergency' were those of Scotland and Wales at the end of April 2019, closely followed by the UK Parliament and many other nations, cities and councils around the world. It is encouraging that many nations and organisations now acknowledge the scale of the problem, even though efforts to reduce emissions remain inadequate. Setbacks have occurred because a number of powerful world leaders in recent years have denied the

▲ **Nordvestfjord, Northeast Greenland National Park, 2018**
Some of nature's finest architecture is displayed in the beautiful icebergs calved from the Greenland Ice Sheet via numerous outlet glaciers. The glacier which delivered this iceberg is the massive Daugaard-Jensen Gletscher which is located at the head of Nordvestfjord.

reality of climate change, and have failed to act, even trying to scupper the rest of the world's attempts to achieve consensus. Unfortunately, global issues such as the coronavirus pandemic and the Russian invasion of Ukraine, have diverted resources and energy away from tackling climate change, while the fossil fuel industry continues to push for further growth of that sector.

CHANGES TO THE CRYOSPHERE AND GLOBAL IMPACTS

What has all this got to do with snow and ice? Well, by reflecting a large amount of solar radiation back into space, the cryosphere moderates Earth's climate, preventing it from over-heating. However, if we enhance the greenhouse effect by adding polluting gases into the atmosphere or by destroying forests, we are heating the planet further. Throughout the previous chapters, the constant theme has been the vulnerability of snow and ice to human-induced global heating, or 'anthropogenic global warming' as it is commonly known. We use the term 'heating' because this better conveys the severity of the state our planet is in. The following examples illustrate this point.

Seasonal snow, river ice and lake ice are all being lost gradually from most parts of the world, shortening the period each year during which solar radiation is reflected back into space. In temperate regions of Europe, from Switzerland to the UK and Scandinavia, winter snow has become much more erratic, unbalancing local economies that rely on winter sports, and altering the ecology of those countries. Pristine areas such as the Arctic are changing even faster. For example, mean annual temperatures in Svalbard have increased a remarkable 6°C (10.8°F) in the last 100 years, and as much as 4°C (7.2°F) in just the last thirty. The summer season is lengthening, disrupting the growing season over the tundra. There is also an increasing frequency of warm rainy spells in winter, which leads to the formation of ice layers on the ground when refreezing takes place. Consequently, grazing animals, such as reindeer, can have difficulty reaching food and be susceptible to starvation. The entire terrestrial ecosystem in the Arctic is thus being subject to unprecedented stress.

Offshore, the sea ice of the Arctic Ocean and surrounding regions is shrinking at a rate of 13% per decade, and it is likely that the permanent ice around the North Pole will start disappearing each summer within a few decades. The consequences for indigenous Arctic peoples, such as the Inuit in Greenland and Canada are already being felt in reduced travelling opportunities and safety over the winter sea ice, which they need for hunting. Conversely, the reduction in sea ice is opening up seaways in the Arctic Ocean for shipping and exploration of hydrocarbons. For example, the Northern Sea Route or Northeast Passage, now provides a reliable route for ships (with icebreaker support) between Europe and eastern Asia.

Changes in the Arctic are also having a growing impact on heavily populated regions of Europe and North America, as increasing heat input to the polar oceans is changing the configuration of the jet stream, causing unprecedented weather events, including heat waves, droughts, torrential rain and freak snow storms. Furthermore, the warming seas mean that whole ecosystems are changing, as ice-loving micro-organisms, on which the food web depends, can no longer survive. Iconic polar species, such as the polar bear, are thus vulnerable. In Antarctica, the sea ice trend, which until recently was erratic, underwent an unprecedented reduction in area in 2023. Here, too, warming seas are affecting ecosystems on which bird species such as Adélie and emperor penguins rely.

If anything, the changes affecting glaciers and ice sheets are even more severe than for sea ice. We know that approximately 65 metres (213 ft) of potential sea-level rise is locked up in land ice, with 7 metres (23 ft) in the Greenland Ice Sheet and 58 metres (190 ft) in the Antarctic Ice Sheet. Mountain glaciers provide around half a metre (1½ ft) and already, whatever we

▲ Ablation Lake, Alexander Island, southern Antarctic Peninsula, 2012

Geologists investigating the past record of glaciation on Alexander Island. The granite boulders record the time when the Antarctic Peninsula Ice Sheet was larger and carried them across a narrow sound from the mainland to the island. Deciphering how the ice sheet behaved in the past gives clues to its response in the future – 'the past is the key to the future'.

do to mitigate the effects of climate change now, many will disappear by 2100. For example, glaciologists have determined on current trends that the Alpine glaciers will lose 50% of their volume from now until 2050, and 95% by 2100. Even if we were able to stop heating the planet today by some remarkable global act of political will, we would still lose 40% by 2050, due to the lag between glacier recession and changing climate. A similar picture of glacier recession is evident in Glacier National Park in the Rocky Mountains, USA. From a count of 35 glaciers in 1966, there was a reduction of 9 by 2015, representing a decrease in area of 39%. The trend has continued since, and by 2100 only small patches are likely to remain. Similar patterns are replicated across the planet.

The great ice sheets are survivors from the last ice age, and they will need thousands of years to disappear completely because their response time is relatively slow. However, since they contain so much more ice than all other glaciers combined, small percentage volume changes have a greater impact on the planet than all of the smaller glaciers. These ice sheets are now discharging ice and meltwater into the ocean at greater rates than at any time in human history, and the trend is accelerating.

The biggest impact of the cryosphere on humanity is in terms of sea-level rise globally, and from a reduction of water resources in mountain regions and surrounding areas. From sea-level rise it is estimated by the end of the century some 800 million people could be adversely affected, and, from declining water resources, over a billion.

A Farewell to Snow & Ice

A FAREWELL TO SNOW AND ICE

Commemorative events have been held recently to highlight the plight of the cryosphere in Switzerland and Iceland. A 'funeral march' and ceremony were held on 22nd September 2019 to mark the disappearance of the Swiss glacier, Pizolgletscher, in northeast Switzerland. It is one of many glaciers that have disappeared or have shrunk drastically in the Alps in recent decades.

In western Iceland, on 18th August 2019, the death of the glacier Okjökull (Ok Glacier) was marked by the unveiling of a memorial plaque, attended by around a hundred people. The event was to draw attention to what is being lost at an accelerating rate, and also a call to action to reduce greenhouse gas emissions. As the Icelandic Prime Minister, Katrín Jakobsdóttir, said at the time: "We see the consequences of the climate crisis. We have no time to lose." The message on the plaque presented in Icelandic and English, and referring to the current concentration of carbon dioxide in the atmosphere reads:

A letter to the future

> *Ok is the first Icelandic glacier to lose its status as a glacier.*
> *In the next 200 years, all our glaciers are expected to follow the same path.*
> *This monument is to acknowledge that we know*
> *what is happening and what needs to be done.*
> *Only you know if we did it.*
> *415ppm CO_2*

This is a stark message that applies to all snow and ice, and hints at what we must do to minimise the consequences of global heating. We have already seen a loss of regular snow, lake ice and river ice in many parts of the Northern Hemisphere, and now we are saying our farewell to increasing numbers of glaciers around the world. How quickly will the snow and ice disappear? That all depends on the actions we take, or fail to take.

◀ **Barentsøya, Svalbard**
The consequences of the decline in sea ice in the Arctic are all too evident for this starving polar bear, as it pauses on the shore of Barentsøya, Svalbard. The species is classified as 'Vulnerable' in the Red List of the International Union for Conservation of Nature' (IUCN). Bears need extensive areas of sea ice from which to hunt seals.

▶ **Ossian Sarsfjellet, Kongsfjorden, Svalbard**
The reindeer of northern Europe and caribou of North America are the same species, and are also classified as 'Vulnerable' by IUCN. In Svalbard, reindeer are increasingly under threat from winter rainfall events. Such events are followed by freezing of wet ground, thus sealing off the mosses and lichens on which the reindeer rely.

◯ **Scoresby Sund, East Greenland, 2018**
Icebergs derived from the Greenland Ice Sheet on the southern side of inner Scoresby Sund in East Greenland during a September sunrise. As Earth's climate heats up, iceberg production is increasing in some regions, such as Greenland and Antarctica, resulting in an acceleration of sea-level rise, thereby having an impact on coastal communities world-wide.

INDEX

Geographical location names in **bold**
Page numbers in **bold** refer to illustrations

Ablation   81, 85, 146, 147, 157
Ablation area   80, 81, 85, 104, 105, 146
Ablation Lake, Antarctica   54, **125**, 247
Accumulation   22, 24, 26, 28, 80, 81, 85, 87, 92, 121, 160, 235
Accumulation area   **80**, 81, 102, 104, 106, 107, 116, 146, **156**
Alaska, USA   1, 10, **19**, 55, **82-83**, 104, 115, **144**, 148, **149**, 150, **156**, 160, **161**, 167, 209, **227**, 230
Alaska Range, Alaska   1, **144**, **156**, 227
Alexander Island, Antarctica   54, **81**, 124, 247
Alpefjord Northeast Greenland National Park, 97
Alpine arnica   **215**
Alpine bearberry   **216**
Alpine flowers   **214-215**, **216**, 223
Alpine mouse-ear chickweed   223
Alps, the   6, **78-79**, 80, 85, 97, 102, 104, 108, **110-111**, 133, 140, 147, 150, **180**, 211, **233**, 234, 235, 237, 241, 251
Alpine snowbell   223
Alpine toadflax   223
Andes, the   81, 129, 133, 146, 233, 237
Antarctica   10, 13, **27**, **34**, **35**, 46, 54, 62, 63, **65-66**, 67, 69, 72, 73, 76, 80, 92, **93-94**, 105, 108, **121**, 124, 160, **161**, 163, 167, 204, **208**, **209**, 246
Antarctic Ice Sheet   76, 102, 105, 108, 160, 161, **172**, 204, 246
Antarctic Peninsula   54, 67, 72, 81, **92-93**, **94-95**, 105, 108, **121**, **125**, 160, **161**, 163, 167, **172**, **173**, **174**, **205**, **208**, **209**, **210**, **212**, 238, **243**, 247
An Teallach, Northwest Highlands, Scotland   187
Aoraki / Mt Cook, Southern Alps, New Zealand   **86**, **88**, **157-158**, 234
Arctic Finland   13
Arctic hare   14, 209, **219**
Arctic flowers   209, **214-215**
Arctic fox   **204**, 209, **221**
Arctic mouse-eared chickweed   **214-215**
Arctic Ocean   60, **61**, 62, 63, **66**, **67**, 69, 108, 208, 209, 210, **218**, 238, 246
Arctic poppy   **214-215**
Arctic, the   10, 13, 22, 28, 60, **61**, 62, 63, **67**, 69, 81, **164-165**, 167, 204, 208, 209, 210, 212, **214**, **218**, **221**, 234, 238, 246, **248**
Arctic willow   204, 209, **216-217**
Asia   104, 133, 230, 246
Astro Lake, Axel Heiberg Island, Canada   8
Atlantic Ocean   63, **202**, **206-207**
Atmosphere   6, 10, 13, 14, 22, 28, 38, **39**, 46, 76, 150, 160, 173, 212, 244, 246, 249
Austria   54, **147**, **233**, 234

Austre Brøggerbreen, NW Spitsbergen, Svalbard, Norway   96, **133**
Avalanche   6, 26, **29**, 87, 157, 235, 237, 238, 241
Axel Heiberg Island, Nunavut, Canada   8, **14**, 75, **84-85**, 87, **104**, **106**, **130-131**, **139**, **142-143**, **170-171**, **214**, **215**, 221

Barentsøya, Svalbard   248
Bäretswil, Switzerland   39
Baltic Sea   63, 69, **73**
Barnes Ice Cap, Baffin Island, Nunavut, Canada   87
Basal sliding   85
Bedrock   6, 13, **84-85**, 150, **161**, 174, 182, **190-191**, **192**, **193**, **194**, **198**, 237
Bergschrund   92
Bergy bits   160, 167, **168**, **170-171**, **176**, **178-179**, 239
Biosphere   10, 212, 244
Bjorne Oer, East Greenland   168
Black ice   46, **47**, 51
Blencathra, English Lake District   **34**, 35
Blizzard   24, 26, 124
Braided river   28, 54, 81, 189, **199**
Breiðamerkurjökull, Iceland   87, **135**
Breithorn, Switzerland   **78-79**
Brungrat, Switzerland, Berner Oberland, 188
British Channel, Franz Josef Land, Russia   61, 63, 67, **70-71**, 234
Bryce Canyon, Utah, USA   **32-33**
Buckskin Glacier, Alaska   156
Bummocks   60
Bylot Island, Nunavut, Canada   55, 107, **123**, **132**, **152-153**, 216

Cadair Idris, Wales   11
Calving   7, 13, 80, **93-94**, 114, **115**, 160, **161**, 163, 167, 174, 176, **177**, 234
Canada   8, **14**, 46, 50, 54, **55**, 62, 63, 68, 74, **84-85**, **86**, 87, **104**, **105**, **106**, **107**, **123**, **126-127**, **130-131**, **132**, **139**, **142-143**, **152-153**, **170-171**, 204, **214**, **215**, **216**, **221**, 230, 234, 246
Canton St Gallen, Switzerland   29
Canton Zürich, Switzerland   **12**, **16-17**, **39**, **40**, 50, **52-53**, **196-197**
Canyons (ice)   13, 129, **132**
Caribou   **14**, 204, 209, **251**
Castlerigg, Keswick, English Lake District   **30-31**
Channels   13, 54, 81, **84-85**, 85, **96**, 102, 124, 129, **126-127**, **135**, 182, 189, **199**, 233
Chile   160, 234
Cierva Cove, Antarctic Peninsula   205
Climate change   4, 7, 10, 13, 81, 55, 105, 62, 76, 81, 102, 104, 105, 115, 117, 124, 229, 234, 236, 244, 246, 247
Climate crisis   7, 212, 251

Climate emergency   7, 244
Clouds, cirrostratus   **12**, 38
Clouds, cirrus   6, 38, **39**, **40**
Clouds, cumulonimbus   6, 38
Clouds, cumulus   38
Cobweb house-leek   223
Coire Ardair, Grampian Highlands, Scotland   **48-49**, **194-195**
Coire Toll an Lochain, An Teallach, Scotland   187
Color Lake, Axel Heiberg Island, Nunavut, Canada   221
Comfortlessbreen, Spitsbergen, Norway   115
Conwaybreen, Spitsbergen   **154-155**, 176
Cornices   **34**, **35**
Corrieschalloch Gorge, Scotland   193
Crevasses   77, 80, 85, **86**, **92-93**, **94-95**, 97, 102, 116, 124, 129, **134**, 146, 150, 160, 163, **164-165**, **166**, 167, 240
Cross-country skiing   234
Crusoe Glacier, Axel Heiberg Island, Canada   75, **104**, **130-131**
Cryosphere   6, 7, 10, 14, 28, 46, 237, 238, 246, 247, 249
Curling   233, 234
Cuverville Island, Antarctic Peninsula   212

Denali, Alaska, USA   1, **144**
Derwentwater, English Lake District   56, **185**
Diamond dust   22
Dickonsfjord, Northeast Greenland   186
Dog-sledging   234

Earth   6, 10, 13, 22, 28, 38, 46, 50, 76, 81, 85, 124, 146, 182, 189, 201, 204, 205, 210, 234, 237, 244, 246, 249
Ella Ø, Northeast Greenland   **219**, **220**
English Lake District   **2-3**, **30-31**, **34**, **35**, **41**, 56, **185**
Eqip Sermia, West Greenland   77, **140**
Erosion   13, 28, 120, 133, **173**, 181, 182, 192
Europe   22, 40, 46, 55, 76, 105, 182, 189, 234, 235, 246, 249
European Alps, the   6, 80, 102, 104, 211, 233
Eyjafjallajökull, Iceland   76, **87**, 157
Firn   76

Finland   13, 21, 24, 25, 35, 36, 40, **42-43**, 46, 69, 73, 233, 234
Fischenthal, Switzerland   45
Fjørtende Julibreen, Spitsbergen, Svalbard, Norway   117
Floes   46, 50, **52-53**, 58, 60, **62**, 63, 67, **170-171**, 208, **218**
Flooding   28, 54, 108, 129, 133, 235, 237, 238, 241, 244

Flyvefjord, Northeast Greenland 61, 228
Fog 24, 40
Folds 85, 97, 150
Foliation 85, **96**, 124, 129, **177**
Fossil fuels 7, 14, 76, 246
Fountain Glacier, Bylot Island, Canada 55, 107, **123**, **132**, **152-153**
Fox Glacier, Southern Alps, New Zealand 94-95, 134
France 76, **89**, **98-99**, **156**, 230
Franz Josef Glacier, Southern Alps, New Zealand 241
Frazil ice 50, 60
French Alps, the **97**, **234**
Frost 38, 43, 44, 124, 230

Garvellach Islands, Argyll, Scotland 200, 201
Gåsefjord, East Greenland 159, **239**
Geosphere 10
George VI Ice Shelf, Antarctic Peninsula 54, 81, **125**
Gerlache Strait, Antarctic Peninsula 121, 173, 174, 208, 210
Germany 43
Glacial trough **183-184**
Glacier **1**, **2-3**, 6, 7, **8**, 10, **11**, 13, **18**, **19**, 22, 28, 46, 54, **55**, 60, **68**, **72**, **75**, 76, 77, **78-79**, 80, 81, **84**, 85, **86**, **87**, **88**, **89**, **92-93**, **96**, 97, **98-99**, **100**, 102, **103**, **104**, **105**, **106**, **107**, **108**, **109**, **110-111**, **112-113**, **114**, **115**, **116**, **117**, **118-119**, **120**, **123**, 124, **125**, **126-127**, **128**, **129**, **130-131**, **132**, 133, **134**, **135**, **136-137**, **138**, **139**, **140-141**, **142-143**, **144**, 146, **147**, **148**, **149**, **150**, 151, **152-153**, **154-155**, **156**, **157**, **159**, 160, **161**, **162-163**, **164**, **165**, **166**, 167, **170-171**, **175**, 182, **183**, 186, **187**, 189, **190-191**, 194, 198, 199, 204, **206-207**, 208, 209, 210, 211, **213**, **224-225**, **228**, 230, **233**, 234, 235, 237, **238**, 240, **243**, 244, **245**, 246, 247, 249
Glacier, bed 13, 80, 85, 139, 150, 152, 189, 235
Glacier, cirque **11**, 50, 76, **89**, 182, 187, **188**, **194-195**
Glacier, hanging 76, **86**, **206-207**
Glacier, ice 76, 80, 87, 124, **148**, 150, 168, 182, 189, 208, 210, 225, 230, 233, 235, 238
Glacier, valley 76
Glacier, niche 76
Glacier, outlet 46, 69, **84-85**, 87, 105, **118-119**, 129, 135, 140, 158, 167, 186, 229, 245
Glacier, polythermal 81, 85, 150
Glacier, warm 80
Glacier, 'mass balance' 81, 85, 102, 104, 150
Glacier, 'snout' 81, 85, 87, **100**, 102, **103**, 104, 115, **138**, **144**, 146, **148**, 182, 189, **240**
Glacier, temperate 80, 81, 85, **134**, **135**, **139**, 150, **182**
Glacier, 'toe' 81
Glacier d'Argentiére, Mont Blanc Massif, France **89**, **156**
Glacier de Trient, Mont Blanc Massif, France 230
Glacier du Chardonnet, Argentiére, Mont Blanc Massif, France 89
Glencoe, Grampian Highlands, Scotland 183-184

Global warming 7, 160, 246
Global heating 7, 10, 14, 28, 46, 69, 85, 102, 210, 212, 230, 246, 246, 251
Gold Harbour, South Georgia, **206-207**
Grease ice 60
Great Lakes, Canada and USA 24, 46
Greenhouse gases 14, 76, 246
Greenland 13, **15**, 28, **61**, 63, **68**, **69**, 76, **80**, **97**, 102, 105, **118-119**, 129, 160, **140**, **159**, 160, 163, **164-165**, **166**, 167, **168**, **169**, **170-171**, 186, 208, 209, **216-217**, 219, 220, 228, 230, 234, **239**, **245**, 246, **250-251**
Greenland Ice Sheet 28, 69, **77**, 80, 102, 105, **118-119**, 129, **140**, **159**, 163, 228, 245, 246
Griessen, Germany 43
Grindelwald, Berner Oberland, Switzerland 102, **224**, **232**, 235
Grosser Aletschgletscher, Berner Alpen, Switzerland 81
Gorge 81, **110-111**, 182, **192**, **193**
Gornergletscher, Valais, Switzerland **78-79**, **140-141**
Graupel 22
Gully Gletscher, Alpefjord, Northeast Greenland 96

Hail 6, 10, 22, 38, **40**
Hairy alpenrose **223**
Haloes 6, 38, **39**, **40**
Hansbreen, Spitsbergen, Svalbard, Norway 162-163
Hardinger Icefield, Alaska, USA 82-83
Helsinki, Finland 73
Hemmental, Switzerland 43
High Arctic 14, 54, **75**, 80, **84-85**, **104**, 115, **170-171**, 204, 209, **214**, **216-217**, 219
Highland icefield 76, **82-83**, 120,
Himalaya, the 81, **109**, 129, 133, 146, 230, 237
Hoarfrost 10, 40, **43**
Hodge Glacier, Northwest Greenland **170-171**
Hornsund, Northwest Spitsbergen, Norway 162, **211**
Hummock 60, **67**, 148, 189, **190-191**, **194-195**
Hydrosphere 10
Hydroelectric power 13, 14, 85, 102, 104, 133, 233

Ice age 13, 19, 76, 102, 108, 160, 183, 185, 189, 192, 194, 225, 230, 237, 247
Ice aprons 76, **86**, **107**
Icebergs 10, 13, 60, **68**, 76, 70, 80, 105, **118-119**, **142-143**, 160, **162**, 163, **164-165**, **166**, 167, **168**, **169**, **172**, **173**, **176**, **177**, **178-179**, 205, 208, **218**, **228**, 234, **245**
Icebergs, tabular 76, 160, 163, **164-165**, 167, **170-171**
Ice caps 28, 76, 80, 81, **87**, 105, 135, 146, 167, **218**
Ice cliffs 60, 80, 160, 163, 167
Ice crystals 6, **12**, **36**, 38, **41**, 46, **50**, 60, 76, **87**, 124
Iceland 6, 63, 80, **87**, 124, 129, **135**, 150, 189, **213**, 235, 237, 249
Ice-roads 10, 50
Ice sheets 6, 10, 13, 76, 102, 105, 108, 121, 133, 146, 160, 182, 189, 204, 237, 244, 246, 247
Ice shelf 46, 54, **81**
Ice skating 234

Icicles **16-17**, **45**, 54, **125**, **172**
Ingmikertikajik, Scoresby Sund, East Greenland 216-217
Iqaluit, Baffin Island, Canada 62
Irrigation 14, 85, 104, 133, 233

Jura Mountains, Switzerland 230, **231**

Kennicot Glacier, Wrangell Mountains, Alaska, USA 148
Keswick, English Lake District **30-31**, **185**
Kongsbreen, Spitsbergen, Svalbard, Norway **114**, **115**, 148, 151, **154-155**
Kongsfjorden, Spitsbergen, Svalbard, Norway 151, **175**, **177**, **214**, **215**, 251
Kronebreen, Spitsbergen, Svalbard, Norway 175

Lake ice 8, 10, 46, 55, 230, 244, 246, 249
Lake Mathieson, Southern Alps, New Zealand 88
Lej da Silvaplauna, Engadin, Switzerland 47, 51, **232**
Lillliehookbreen, Spitsbergen, Svalbard, Norway **178-179**
Liverpool Land, East Greenland 15
Longyearbyen, Svalbard 214
London, England 230
Luosto, Lapland, Finland 12, 21, **42-43**, 232
Lyskamm, Switzerland **78-79**

Magga Dan Gletscher, East Greenland 159
Martigny, Switzerland 230
Matterhorn, Switzerland **78-79**, **180**, 235
Mauna Loa, Hawaii, USA 244
McMurdo Ice Shelf, Ross Sea, Antarctica 27, 35, **64-65**, 73
Meares Glacier, Alaska **161**
Melt-streams 81, 28, 81, 85
Meltwater 6, 13, 19, 28, 46, **58**, 81, 85, 102, 104, 116, 124, **126-127**, 129, 133, **134**, 135, **136-137**, **139**, **140**, 150, 160, 182, 189, 192, **199**, 204, 230, 235, 237, 247
Mer de Glace, Mont Blanc, Chamonix, France 98-99, 234
Minna Bluff, Ross Sea, Antarctica 34
Monlesi ice caves, Jura Mountains, Switzerland 231
Monte Rosa, Switzerland **78-79**
Moraine 4, **18**, 76, 102, **109**, 129, **144**, 146, 149, 150, 182, 189, **194-195**, 237
Morteratsch Valley, Switzerland **18**, **128**, **222**
Moss campion **214**, **223**
Moulins 85, 129, 150, 182
Mount Gaudry, Adelaide Island, Antarctica **72**
Mount Steele, Yukon, Canada 86
Mount Tasman, Southern Alps, New Zealand 88
Mudflows 235, 237, 238
Multi-year ice **58**, 60, 62, 67, 69, 73, 204, 209
Musk ox 14, 209, **220**

Nabesna Glacier, Wrangell Mountains, Alaska, USA 149
Nepal 108, **109**, 237
Nevada del Ruiz, Colombia 235
Nevado Huascaran Norte, Peru 237
New Zealand Southern Alps 80, **86**, **88**, **94-95**, 133, **134**, **139**, 146, **157**, 182, **199**, 234, **240**
Nilas 60, **61**, 68
Nordaustlandet, Svalbard, Norway 167, **218**, **250**
Nordvestfjord, East Greenland 61, 68, **118-119**, **164-165**, **166**, 167, **168**, **169**, 245
North America 1, 22, 24, 28, 40, 46, 76, 87, 133, **144**, 146, 182, 209, 211, 212, **219**, 234, 246, 251
North America, Western Cordillera 80, 133, 146, 234
North Pole 59 63, 209, 234, 246
Northern Scandinavia 40, 230
Norway 6, 27, 96, 104, **115**, **116**, **117**, **120**, **133**, **148**, **151**, **153-154**, **162**, **167**, **175**, **176**, **178-179**, 182, 208, 209, **211**, **214**, **215**, **218**, **221**, 230, **234**, **236**, 246, **248**, **249**
Nunavut, Canada 8, 14, 68, 75, **84-85**, 87, **104**, **105**, **106**, **107**, **123**, **126-127**, **130-131**, **132**, **139**, **142-143**, **152-153**, **170-171**, **214**, **215**, **216**, **221**
Ny-Ålesund, Northwest Spitsbergen, Svalbard, Norway 221

Oberaargletscher, Berner Oberland, Switzerland 138
Obers Ischmeer, Berner Oberland, Switzerland **224-225**
Ogives 85, **98-99**
Okjökull, Iceland 251
Orne Harbour, Antarctic Peninsula 238
Ossian Sarsfjellet, Spitsbergen, Svalbard, Norway 249

Pack ice 60, 62
Palmer Land, Antarctic Peninsula 92
Pancake ice 60
Pasterzenkees, Hohe Tauern National Park, Austria 147
Paradise Harbour, Antarctic Peninsula **90-91**, **121**, **173**, **174**
Paris, France 230
Penguins 13, **203**, 204, **206-207**, **208**, **209**, 212, 234, 246
Permafrost 204, 208, 211, 212, 226, 244
Permanent ice 10, 46, 246
Peru 133, 233, 237
Peruvian Andes 237
Pfaffikersee, Canton Zurich, Switzerland 50
Pizolgletscher, Switzerland 251
Polar bear 13, **15**, **67**, 204, 210, 212, **218-219**, 234, **248**, 250
Polynyas 60
Pontresina, Engadin, Switzerland 108, **241**
Pressure ridges 46, 54, 60, 63, **66**
Purple saxifrage **214**

Rees River, Southern Alps, New Zealand 199
Reindeer 13, 14, 209, 230, **232**, 246, **250**

Reindeer sledging 234
Resolute, Cornwallis Island, Canada 68
Rhonegletscher, Furka Pass, Switzerland **194**, 234
Rime 10, **13**, 38, 40, **41**, **42-43**
River ice 6, 10, 50, 55, 244, 246, 249
River Thames, London, England 230
Rob Roy Glacier, Southern Alps, New Zealand **139**
Roches moutonnées 182
Rocky Mountains, USA 247
Rovaniemi, Finland 40
Russian Arctic 50

Saariselka, Lapland, Finland 24, 25, 35
Sastrugi 28, **34**, **35**
Scafell Pike, English Lake District **41**
Scandinavia 40, 80, 230, 234, 235, 246
Scheuzer's bellflower or Campanaula **223**
Scoresby Sund, East Greenland **250-251**
Scotland **48-49**, **183-184**, **187**, **190-191**, **193**, **194-195**, **200**, **201**, 244, 246
Scott Base, Antarctica 34, **64-65**, 73
Sea ice 10, 14, 59, 60, 62, **63**, **64-65**, 67, 69, **73**, **170-171**, 204, 208, 209, 210, 212, 218, 234, 237, 238, 244, 246, 249
Sediment 13, 50, 81, **123**, 129, **133**, **139**, 140, 146, **147**, **149**, 150, **151**, **170-171**, 182, 198, 201, 209, 211
Sedov Station, Franz Josef Land, Russia **70-71**
Sérac 85
Sefstrøm Gletscher, Alpefjord 96
Shore-fast ice 60
Siberia 10, 46, 54, 76, 182, 230
Skalafellsjökull, Iceland 213
Skiing 10, 85, 233, 234
Sledging 10, 230, 234
Sleet 22, 38
Slovakia 54
Smeerenburgbreen, Spitsbergen 120
Snowbanks 28
Snowboarding 234
Snow bridge 85
Snowdonia 11
Snow dunes 28, **34**
Snowfall 10, 11, 22, 24, 26, 35, 46, 76, 80, 87, 104, 105, 146, 230, 235
Snowfields 6, 106, 199
Snowflakes 6, 22, 76
Snow flurries 24
Snowpack 22, 26, 28, 35, 124
Snow scallops 28, **34**
Snow-shoeing 234
Snow showers 24
Snow storms 24
Spring melt 28, 54
Søbotn, Troms 236
Sodankylä, Finland 36
Solothurn, Switzerland 198
Southern Axel Heiberg Island, Canada **84-85**
South Croker Bay Glacier, Devon Island, Canada 105
South Georgia 160, **203**, **206-207**
Southern Ocean 67, 108, 160, 205, 208
Spring gentian 223

Suilven, Northwest Highlands, Scotland **190-191**
Stalagmites 54
St Andrew's Bay, South Georgia 204
Stratification 26, 46, 87
Steel Glacier, Yukon, Canada 86
Striations 182
Sun dogs 38, **40**
Supercooled water 22, 38, 40, 43, 50
Svalbard 96, 104, **115**, **116**, **117**, **120**, **133**, **148**, **151**, **153-154**, **162**, **167**, **175**, **176**, **178-179**, 208, 209, **211**, **214**, **215**, **218**, **221**, **234**, 246, **248**, **249**
Switzerland 12, 15, 18, 26, 29, 39, 40, 43, 45, 47, 50, 51, **52-53**, **78-79**, 100, 102, 109, **112-113**, **128**, **136-137**, **138**, **140-141**, **147**, **149**, 180, 188, **192**, **194**, **196-197**, **198**, **222**, **223**, **225**, 230, **231**, **232**, 235, **241**, 246, 249

Tasman Glacier, Southern Alps, New Zealand **157**
Thermal stratification 46
Thompson Glacier, Axel Heiberg Island, Canada 8, **126-127**, **139**, **142-143**
Threlkeld, English Lake District 4
Tibetan Plateau, China 55
Tokositna Glacier, Alaska, USA 1, **144**
Troms, Norway 27, 236
Tunnels 13, 81, 124, 129, 133, **136-137**, **138**, 182, 233, 234, 237

UK 24, 40, 108, 182, 230, 244, 246
Ukraine 244
Underground ice 10
Unteraargletscher, Berner Oberland, Switzerland 103, **147**
Unterer Grindelwaldgletscher, Switzerland **192**, 235
USA 33-34, 46, 50, 104, 182, **226**, **227**, 247

Vadrec del Forno, Maloja, Switzerland 149
Vadret del Forno, Canton Graubünden, Switzerland 100
Vadret da Morteratsch, Engadin, Switzerland 18, **112-113**, 129, **136-137**, **223**, 225
Vadret de Tschierva, Pontresina, Switzerland **109**

Wahlenbergbreen, Spitsbergen, Svalbard, Norway 116
Walenstadt, Canton St Gallen, Switzerland 29
Weissfluhjoch, Davos, Switzerland 26
White-out 24, 26
Wormald Glacier, Antarctic Peninsula 243
Wrangell Mountains, Southern Alaska 19, 149

Yellow mountain saxifrage 223
Yellowstone National Park, Wyoming, USA 226
Yukon, Canada 10, 86, 104, **115**

ACKNOWLEDGEMENTS

This book has been a retirement project, and follows half a century of glaciological and geological research in the polar regions and various mountain ranges around the world. Over this period, we have benefitted from collaboration with colleagues, too numerous to mention, from many different countries.

We have been provided with the opportunities to undertake our glaciological research and photography by our various employers: the University of Manchester, the Swiss Federal Institute of Technology (ETH) in Zürich, the University of Cambridge, Liverpool John Moores University, Aberystwyth University (Wales), and Kantonsschule Zürcher Unterland (Bülach, Switzerland).

In addition, we have benefitted from collaborative projects with the Victoria University of Wellington and the University of Otago in New Zealand; Queen's University at Kingston, Ontario, the University of British Columbia and the University of Alberta in Canada; the University of Ohio in the USA; as well with numerous other institutions in the UK and Switzerland.

Research funding and logistical support has come from many sources, notably the UK Natural Environment Research Council, the British Antarctic Survey, the Alfred Wegener Institute for Polar and Marine Research (Germany), the Polar Continental Shelf Program (Canada), Antarctica, New Zealand, and the Australian National Antarctic Research Expeditions.

MH is also grateful for the opportunity to work with polar tour-operator, Quark Expeditions, which has enabled him to take many of the photographs illustrated in the book.

***

JA is particularly grateful for the support and understanding of his wife, Pamela Alean-Kirkpatrick, for his glaciological enterprises.

Many of the photographs could not have been taken but for the skills of the pilots of helicopters and fixed-wing aircraft, the boat crews of both large and small research vessels, and the field assistants/mountaineers who have kept us safe.

Thanks are due to Dr Ailsa Benson and Dr Diana Mitchell, who have reviewed and read all or part of the book from the perspective of the non-glaciologist, ensuring that the text is not too technical. Also, Professor Terence Sloan, who advised on the over-arching theme in this book, namely, climate change.

Finally, we appreciate the efforts of our publisher, Alexandra Papadakis, and her design and editorial team, in creating a beautiful book that shows our photographs to their best advantage.

Michael Hambrey and Jürg Alean
September, 2024, Threlkeld and Aberystwyth, UK,
and Eglisau, Switzerland

ABOUT THE AUTHORS

The authors standing on a glacier table, Vadret Pers, Engadin, Switzerland.

Michael Hambrey (left) is Emeritus Professor of Glaciology and former Director of the Centre for Glaciology in the Department of Geography & Earth Sciences at Aberystwyth University, Wales, UK. He was also founding director of the Climate Change Consortium of Wales. His research interests include glacial geology and structural glaciology. He has published nearly 200 scientific papers and several books, including university-level textbooks *Glacial Environments* (1994) and, with Jürg Alean, *Colour Atlas of Glacial Phenomena* (2017). In addition Michael has co-written popular science books *Glaciers* (2nd edn, 2004) and *Gletscher der Welt* (2013) with Jürg Alean, and *Islands of the Arctic* (2002) and *The Continent of Antarctica* (2018) with Julian Dowdeswell.

Michael has undertaken glaciological fieldwork in Norway, New Zealand, the Swiss Alps, the Andes, the Himalaya, the Canadian Arctic, Yukon, Alaska, Greenland, Svalbard and Antarctica. For his work in the polar regions, he was awarded the 'Polar Medal' twice by the late Queen Elizabeth II (in 1989 and 2012) and the SCAR (Scientific Committee on Antarctic Research) Medal for Scientific Excellence (in 2019). The naming of 'Hambrey Cliffs' on James Ross Island in Antarctica also recognises his contribution to polar science. Michael has served on a number of UK national and international committees dealing with glacial and polar issues, and has been responsible for editing several scientific journals and conference volumes.

Jürg Alean (right) was formerly a teacher of geography at the Kantonsschule Zürcher Unterland in Bülach, Switzerland. He has undertaken extensive fieldwork in the Swiss Alps, the Canadian Arctic, Alaska and South America. His research has led to various scientific papers, in particular concerning dangerous glaciers and ice avalanches. He has also published many popular scientific articles and several books, for example, *Gletscher der Alpen* (2010) and *Gletscher der Welt* (2013), as well as those mentioned above with Michael Hambrey.

Jürg is member of a team of volunteers maintaining 'SwissEduc.ch', a Web platform dedicated to providing teaching materials mainly for secondary education, and also hosting www.glaciers-online.net, where both authors present a wide range of their photographic work on and around glaciers all over the world.

Michael and Jürg's book *Glaciers* (1st edn) earned the Earth Science Publishers (USA) Outstanding Publication Award in 1995.